Marie Berrondo

Translated and edited by
Peter Dax and Mariana Fitzpatrick

Mathematical Games

PRENTICE-HALL, INC., Englewood Cliffs, New Jersey 07632

Library of Congress Cataloging in Publication Data

Berrondo, Marie.
Mathematical games.

Translation of: Les jeux mathématiques d'Eureka.
1. Mathematical recreations. I. Dax, Peter.
II. Fitzpatrick, Mariana. III. Title.
QA95.B46413 1983 793.7'4 82-21540
ISBN 0-13-561399-X
ISBN 0-13-561381-7 (A Reward book: pbk.)

This book is available at a special discount when ordered in bulk quantities. Contact Prentice-Hall, Inc., General Publishing Division, Special Sales, Englewood Cliffs, N.J. 07632.

10 9 8 7 6 5 4 3 2 1

Printed in the United States of America

Editorial/production supervision: Marlys Lehmann
Cover design: Hal Siegel
Manufacturing buyer: Pat Mahoney

ISBN 0-13-561399-X

ISBN 0-13-561381-7 {A REWARD BOOK : PBK.}

Prentice-Hall International, Inc., *London*
Prentice-Hall of Australia Pty. Limited, *Sydney*
Prentice-Hall Canada Inc., *Toronto*
Prentice-Hall of India Private Limited, *New Delhi*
Prentice-Hall of Japan, Inc., *Tokyo*
Prentice-Hall of Southeast Asia Pte. Ltd., *Singapore*
Whitehall Books Limited, *Wellington, New Zealand*
Editora Prentice-Hall do Brasil Ltda., *Rio de Janeiro*

Contents

1. The mysteries of probability

1. Going hunting

Three hunters shoot simultaneously at a hare. The first is known to hit a target 3 times out of 5, the second 3 times out of 10, and the third only once in 10 times. What is the probability that the hare will be wounded by at least one of the hunters?

2. As many boys as girls

Among families with two children, half have as many girls as boys. Is the same true of families with four children? (It is assumed that there is an equal probability of having a boy or a girl at each birth.)

3. Turnpike

Knowing that the probability of having an accident while traveling 1 km of turnpike is p, what is the probability of an accident occurring during a 775-km journey?

4. Park bench

One spring day, three boys and two girls line up on a park bench to enjoy the sun. They choose their seats at random. Is there a greater chance that the two girls will be separated or that they will be seated side by side?

5. White marbles, black marbles

Jim keeps his 10 prize marbles, 5 white and 5 black, in a jar. He draws 3 at random, using two methods. The first time he replaces each marble in the jar before drawing the next. The second time he draws the marbles without replacements. By which method is he more likely to obtain 1 white marble and 2 black ones?

6. War in the Middle Ages

During a grim medieval battle, 85% of the warriors lose an ear, 80% lose an eye, 75% lose an arm, and 70% lose a leg. What is the smallest percentage possible of combatants who will lose all of the above?

7. At the dentist's office

Two women and 10 men wait their turn at a dentist's office. The reception room contains eight copies of the latest news magazine and four copies of the morning paper. How many different ways can this reading matter be distributed, given that both women read the same publication?

8. Shooting dice

One evening Smith and Jones get involved in a dice game. They take turns throwing two dice. When the total shown is 7, Jones wins a point; when the sum shown is 8, Smith scores one. Whom would you bet on to win?

9. Unlucky in love

Three boys and three girls are inseparable. Each of the six is in love with one of the three group members of the opposite sex, chosen at random. One of the girls reports sadly that none of the group is loved by the person he or she cares for. Is this sad state of affairs so surprising?

10. Auto race

The race track is terribly dangerous. Drivers must first cross a very narrow bridge from which 1 out of every 5 cars falls into the water. Next comes a grueling hairpin turn that sends 3 out of every 10 cars into a ditch. A frightful tunnel follows, so dark that 1 out of every 10 cars hits a wall and never reemerges. Last comes a sandy stretch in which 2 out of 5 cars bog down.

Given these pitfalls, what percentage of those cars competing do not make it through the race intact?

11. How many are we?

You tell us the answer, knowing that the probability that at least two of us have birthdays on the same day is less than 1/2, but that this would not be the case were we one more in number.

12. Convoy

Four escort vessels, seven merchant ships, and three aircraft carriers are assembled in an East Coast port during World War II. They will form a convoy, carrying food and munitions to Europe. An escort vessel will lead the convoy, followed by three merchant ships, then a carrier. An escort vessel will take up the rear. In how many ways can this convoy be organized?

13. Town drunk

The police want to find the town drunk. There are four chances in five that he is in one of the town's eight bars, chosen at random. Two officers visit seven bars without success. What are the chances that the drunk will be in the remaining bar?

14. Where is Beatrice?

One pleasant summer day, Frank decides to try to find Beatrice at the shore. Is she on the beach (one chance in two)? At the tennis court (one chance in four)? At the snack bar (one chance in four)? If she is on the beach, which is wide and crowded, Frank has one chance in two of not finding her. If she is at the court, he has one chance out of three of missing her. If she is at the snack bar drinking a cola, he is sure to see her.

Frank looks in all three places without success (it is assumed that Beatrice remains in one spot during his search). What is the probability that Beatrice was on the beach?

15. A sheik and his oil

My sheikdom includes a vast desert, where my palace stands, and some underwater acreage of which I am very fond. My undersea holdings are extensive (as much as a third of my desert) and contain part of my oil field. Knowing that I own three times more square kilometers of desert containing oil than underwater terrain without oil and that one-seventh of my non-oil-bearing sheikdom is undersea, can you tell me exactly what proportion of my oil field lies beneath the water?

16. Pick a card

Take a card from my deck at random, then replace it. Do this three times. In the process, you will have 19 chances in 27 to come up with at least one face card (jack, queen, or king). This is because the proportion of face cards in my deck is . . . Can you complete this sentence?

17. Poll taking

Within a given population, 42% of the people have never skied and 58% have never taken a plane, while 29% have done both. Is it therefore more likely to find a skier among those who have never flown or to find an air passenger among those who know how to ski?

18. Life expectancy

An ethnologist establishes that within a primitive population, 25% of the people die at age 40, 50% at age 50, and 25% at age 60. He decides to pick two natives at random for further study. What is the life expectancy of the native between the two chosen who will live longer?

19. Red light, green light

A broad one-way street contains two consecutive traffic lights, each of which remains green two-thirds of the time. A driver observes that at his usual driving speed, if he makes the first light on green, the second will be green on his arrival three times out of four. What is the probability that the second light will be red on arrival, knowing that our driver went through the first light on red?

20. Flight attendants

Twenty flight attendants apply for a transatlantic flight. Seven are blond, the rest have dark hair. Three are picked for the job at random.

What is the probability that there will be at least one fair-haired and one brunette attendant on board?

21. Sunday paper

Every Sunday morning Mr. Smith takes a walk. Will he remember to pick up the Sunday paper, Mrs. Smith wonders, or should she go out and buy it herself? "Do you think Dad will bring home the paper?" she asks her son. The boy guesses wrong one time out of three. Since it is twice as annoying to have no paper than to have two, what do you think Mrs. Smith should do? Buy a paper whatever her son says, or only purchase one if he says that his father will forget?

22. Raising an army

A warrior king decides that for military reasons he needs more boys than girls in his country. He therefore proclaims that no family may have more than one daughter. As a result, the women of the land stop bearing children once they have had a girl. What proportion of male children will result from this practice?

23. History test

During a history exam, Jim and Joe sit side by side. They are asked to supply two dates: the year that President Jefferson was inaugurated and the year that Louisiana was purchased. Jim knows that the answers are 1801 and 1803, but he doesn't know which date goes with which event. He whispers to Joe for help. Joe has three chances out of four of knowing the right answer, but there is one chance out of four that he will give Jim the wrong reply out of spite. Is it therefore better for Jim to take Joe's word or to guess?

24. Twins

We know that 3 births out of 250 result in twins and that one time out of three the twins will be identical. Given this information, deduce a priori the possibility of a pregnant woman having twins of opposite sexes.

25. Jaws

A seaside resort along the Pacific Ocean is equipped with an electronic shark detection system which sounds an alarm on the average of 1 day out of 30. There are 10 times more false alarms than there are undetected sharks. We also know that only three out of four sharks are detected. Given the above information, what is the percentage of "peaceful" resort days if that term is defined as days unmarred by alarms or sharks?

26. Meteorology

It rains here one day out of three. Our local meteorologists, who are pessimistic by nature, are wrong in their daily forecasts one time out of two when they should have predicted good weather but wrong only one time out of five when they should have forecasted rain.

Each morning, Francine leaves home for the day. If she departs without an umbrella and it rains, she is twice as annoyed as if it had been fair and she had taken her umbrella with her. "Would I be wiser," she wonders, "to listen to the morning weather forecast and take my umbrella only if rain is predicted? Or should I consistently carry it? Or should I never take it?" What would you advise Francine to do?

27. Passenger ship

During a long and circuitous passage, the captain observes that a fourth of the passengers leave the ship and an equivalent number join the ship at each port of call, that of those who leave ship only 1 out of 10 boarded at the preceding port of call, and that the boat is always full. Given these facts, whenever the boat is at sea, what proportion of the passengers will not have boarded it at either of the last two ports of call?

28. Mrs. Martin's umbrella

One Saturday night, two brothers, John and Peter, discuss their neighbor Mrs. Martin. They agree that she goes out once and once only every Sunday, that two times out of three she takes an umbrella, and that even in fine weather she carries her umbrella one time out of two. But while John insists that on rainy days she always takes her umbrella, Peter maintains that she sometimes leaves it behind. Knowing that in Mrs. Martin's neighborhood it rains one day out of two, which of the brothers is right?

29. Parking meter

While visiting Paris, I need to park for 10 minutes to pick up a package but am reluctant to feed the meter at the cost of 2 francs per half-hour. The police pass every 2 hours and the fine is 48 francs. I decide to ask a café owner if the police have been by during the previous 60 minutes. There is one chance out of four that I will get the wrong answer since the café owner may think he saw an officer when he didn't or may have been looking elsewhere when the police, in fact, passed. What would you do in my place, moral issues aside? Automatically pay the 2 francs or do so only if the café owner affirms that the police have not passed during the previous hour?

30. Hugh and Caroline

Hugh and Caroline arrange to meet in front of the Empire State Building at an unspecified time between 11 and noon. Each is to arrive during the given hour but is to leave after 15 minutes if the other has not shown up. What is the probability that Hugh and Caroline will, in fact, meet?

31. Rain or shine?

Statistics show that in a given region it rains one day out of four. We also know that if it has rained one day, it will rain two times out of three the next. What is the probability that it will *not* rain the day following a day of fair weather?

32. Job hunt

On release from the service, Mark looks for work. He writes to various companies that he believes can use his skills. He has one chance in five to be accepted per application. He will not stop sending out résumés until he has at least three chances out of four of finding work. Knowing that the logarithms to the base 10 of 3, 4, and 5 are 0.477, 0.602, and 0.699, respectively, how many letters will Mark have to write?

33. Family meal

Every Sunday, a married couple invites both of their mothers for lunch. Unfortunately, neither husband nor wife gets along with his or her mother-in-law and each knows that there are two chances out of three that they will quarrel with their spouse's mother. Moreover, the uninvolved partner will then join in the argument, taking one side or the other at random: one time out of two turning against his or her mother and the other time quarreling spouse against spouse. Knowing that the arguments each partner has with his or her mother-in-law are unrelated, what is the proportion of Sundays when the couple themselves will not quarrel?

34. Late again?

To get to work, Morris sometimes travels by car (arriving late one time out of two) and sometimes by subway (arriving late only once every four times). If he gets in on time one day, he makes a point of using the same method of transportation the next day. Conversely, on days after he has arrived late, he tries the alternative method. This being the case, how likely is Morris to be late on his 467th day of work?

35. The dailies

In a vacation town, 28% of the adults read the *Sun*, 25% read the *Post*, and 20% read the *Times*. Also, 11% read both the *Sun* and the *Post*, 3% read the *Sun* and the *Times*, 2% read the *Post* and the *Times*, and 42% read none of these papers. What percentage of vacationing adults read all three papers?

36. International gathering

Four Frenchmen, two Africans, three Russians, and four Swedes assemble to discuss forest development. During the opening meeting, the delegates from each country sit next to one another on a bench that seats 13 people. They remain side by side during the succeeding working sessions held around a round table, which also has 13 seats. How many different ways can the delegates be seated in each of these two cases?

37. Sensible or sensitive?

During an English comprehension test, a foreigner is asked the following question: "If Nancy has sense, does this mean that she is sensible or sensitive?" If the student guesses the answer, she has one chance in two of being wrong. She could copy from her neighbor, but she knows from experience that the latter answers such vocabulary questions wrong one out of five times. Besides, there is one chance in ten that the teacher would catch her cheating, which would be three times worse than replying incorrectly. What should the student do, moral issues aside? Copy from her neighbor or choose between the two answers at random?

38. Vacation memories

During my trip to Europe, I saw medieval churches, triumphal arches, waterfalls, and castles. I have pictures of half of these sights. I saw three times more medieval churches than triumphal arches and as many castles as waterfalls. One-fourth of my photos depict castles, although I only snapped one out of every two that I saw. However, I have pictures of all of the triumphal arches that I visited. Can you calculate what proportion of my vacation photo collection depicts triumphal arches?

39. Washing-up time

Ten couples dine together. Five of the 20 people are chosen at random to do the dishes. What is the probability that there will not be a couple among them?

40. Homework

At the end of each class, a French teacher, Madame Dubois, gives her 30 students a page of vocabulary to memorize for the next session. During each class she quizzes 1 student out of 3, 1 out of 10 of whom was also quizzed during the previous session. Knowing this, a lazy student decides to do his homework only if he had not been called on last time. "Will it happen more often that I will have learned my lesson for nothing, or that Madame Dubois will scold me for not doing my homework?" he wonders. What do you think?

41. Twin sisters

Ann and Bridget are identical twins. They sit side by side in the classroom, one always to the left, the other to the right. They both swear that they are Bridget. Their teacher, having observed that in general the twin to the right fibs one time out of four while the twin to the left fibs one time out of five, wonders what the probability is that Ann is the twin to the left. Can you help her find the answer?

42. At Le Mans

One year only three makes of car take part in the 24-hour auto race at Le Mans: Ford, Jaguar, and Maserati. If the odds that Ford loses are 2 to 2, and that Jaguar loses are 5 to 1, what are the odds that Maserati will lose? (In this context "lose" means not come in first.)

Probability debunked

1. Going hunting

If the hare is not wounded, it is because the first hunter missed him (probability 2/5) *and* the second (probability 7/10) *and* the third (probability 9/10).The probability that the hare will not be wounded is therefore $2/5 \times 7/10 \times 9/10 = 126/500$, and the probability that the hare will be wounded is $1 - 126/500 = 0.748$.

2. As many boys as girls

Among families with 4 children, assuming that there is an equal probability of having a boy (B) or a girl (G) at each birth, there are $2 \times 2 \times 2 \times 2 = 2^4 = 16$ possible groupings by gender: *GGGG, GGGB, GGBG, GBGG, BGGG, GGBB, GBGB, BGGB, GBBG, BGBG, BBGG, GBBB, BGBB, BBGB, BBBG, BBBB*. Among these 16 groupings, only 6 have as many girls as they have boys: that is, 6 out of 16 or 3 families out of 8 as opposed to 1 out of 2 families with 2 children.

3. Turnpike

The probability of not having an accident over 1 km is $1 - p$. The probability of not having an accident over 775 km is therefore $(1 - p)^{775}$, and the probability of having an accident over such a distance is $1 - (1 - p)^{775}$.

4. Park bench

Number of ways for this group of three boys and two girls to sit on the five-seater bench:

$$C_5^2 = \frac{5!}{3!\,2!} = 10$$

Number of ways for this same group to sit on the bench with two girls placed side by side = 4 (obvious).

Hence it is more likely for the two girls to find themselves separated than together.

Note: C_n^p represents the number of possible combinations of n objects taken p at a time and is equal to

$$\frac{n!}{p!(n-p)!}$$

where $n! = 1 \cdot 2 \cdot 3 \cdot 4 \cdot \ldots \cdot (n-1) \cdot n$.

5. *White marbles, black marbles*

When using the first method (draw and replace each marble), we have P_1 = (number of ways of having 1 white marble among the 3 selected marbles) × $(1/2)^3 = 3 \times 1/2^3 = 3/8 = 9/24$

When using the second method (draw but do not replace the marbles), we have P_2 = (number of ways of choosing a marble from among 5 × number of ways of choosing 2 marbles from among 5) ÷ number of ways of choosing 3 marbles from among 10, or

$$(C_5^1 \times C_5^2) \div C\ 3/10 = 5/12 = 10/24$$

The desired result is more likely to be obtained by using the second method.

6. *War in the Middle Ages*

Since 85% of the warriors lost an ear and 80% an eye (a total of 165%), at least 65% of them must have lost both an ear and an eye.

Since at least 65% of the warriors lost both an ear and an eye and 75% lost an arm (a total of 140%), at last 40% of them must have lost an ear, an eye, and an arm each.

Since at least 40% of the warriors lost an ear, an eye, and an arm and 70% lost a leg (a total of 110%), at least 10% of them must have lost an ear, an eye, an arm, and a leg each.

(This problem was set by Lewis Carroll.)

7. *At the dentist's office*

If the two women are reading the morning papers: the number of ways to give the remaining two morning papers to the 10 men is:

$$C_{10}^{2} = \frac{10!}{2!\,8!} = 45$$

If the two women are reading news magazines, the number of ways to give the four morning papers to the 10 men is:

$$C_{10}^{4} = \frac{10!}{4!\,6!} = 210$$

Therefore, in all, there are 45 + 210 = 255 possible ways to distribute the newspapers and magazines.

8. *Shooting dice*

Whatever the result of the throw of the first die, Jones knows that he will have one chance in six of winning when he throws the second die (throw a 1 after a 6, a 2 after a 5, etc.); whereas if the result of the throw of the first die is 1, Smith has no chance of winning. Better bet on Jones!

9. *Unlucky in love*

Let B_1, B_2, and B_3 be the three boys and G_1, G_2, and G_3 be the three girls. Let us identify all the possible combinations by means of the tree pictured here.

The total number of combinations corresponding to the sad state of affairs reported by the girl is $3 \times 52 = 156$. The total number of possibilities is $3^6 = 729$. The probability of none of the group being loved by the person they care for is therefore $156/729 = 0.214$, which is not such a rare occurrence (more than 1 chance out of 5).

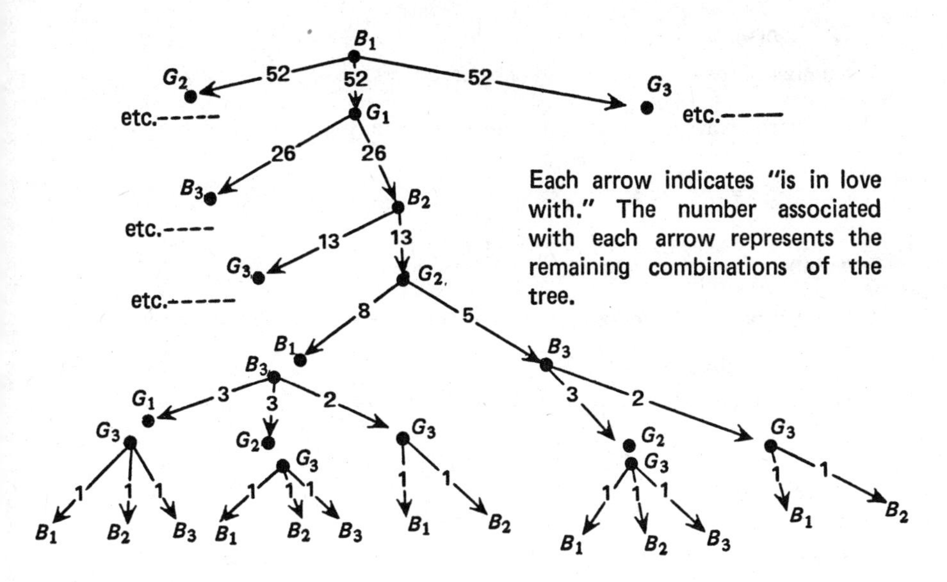

10. Auto race

For a car to survive the course, it is necessary that it not fall off the bridge, nor miss the hairpin turn, nor hit the tunnel wall, nor get bogged down. The possibility of this happening is therefore

$$4/5 \times 7/10 \times 9/10 \times 3/5 = 0.302$$

The percentage of cars that do not make it through the race is therefore 100 − (100 × 0.302) ≃ 70%

11. How many are we?

Probability that two people have different birthdays: 364/365.

Probability that three people have different birthdays: (364/365) · (363/365), etc.

For *n* people

$$P = (364/365) \cdot (363/365) \cdot \ldots \cdot [(365 - n + 1)/365]$$

It can be checked that the product of these fractions becomes less than 1/2 when *n* goes from 22 to 23. We are therefore 22.

12. Convoy

Number of ways of choosing the first escort vessel: 4.

Number of ways of choosing the three merchant ships:

$$C_7^3 = \frac{7!}{3!\,4!} = 35$$

Number of ways of choosing the carriers: 3.

Number of ways of choosing the last escort vessel: 3 (for there are only three left).

Hence the number of ways to organize the convoy is

$$4 \times 35 \times 3 \times 3 = 1260$$

13. Town drunk

A priori, there are 8 chances out of 10 for the drunk to be in one of the town's eight bars, and therefore 1 chance out of 10, a priori, that he will be in the eighth bar. Whereas the probability, a priori, that he will not be found in any of the seven bars visited is:

$$1 - 7/10 = 3/10$$

The probability that the drunk is in the eighth bar, knowing that he is not in any of the seven other bars, is therefore

$$(1/10) \div (3/10) = 1/3$$

The two officers thus have one chance in three of finding the drunk in the eighth bar.

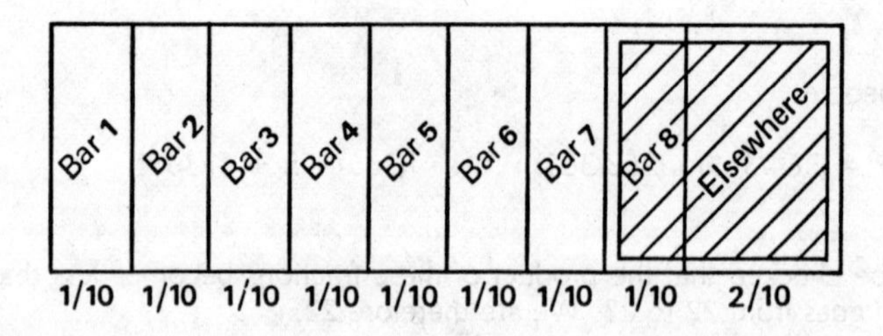

14. Where is Beatrice?

Probability that Beatrice and Frank meet

at the beach: $(1/2) \cdot (1/2) = 1/4$

at the tennis court: $(1/4) \cdot (2/3) = 1/6$

at the snack bar: $(1/4) \cdot (1) = 1/4$

Total: $4/6 = 2/3$.

The probability that they do not meet at all is therefore $1 - 2/3 = 1/3$.

However, the probability that Beatrice and Frank miss each other at the beach is $(1/2) \cdot (1/2) = 1/4$.

The probability that Beatrice was at the beach given that Frank did not see her is therefore

$$(1/4) \div (1/3) = 3/4$$

15. A sheik and his oil

Let S be the total area of my sheikdom.

Area under the sea $= S/4$.

Let x be the ratio of the undersea acreage without oil to the totality of my sheikdom.

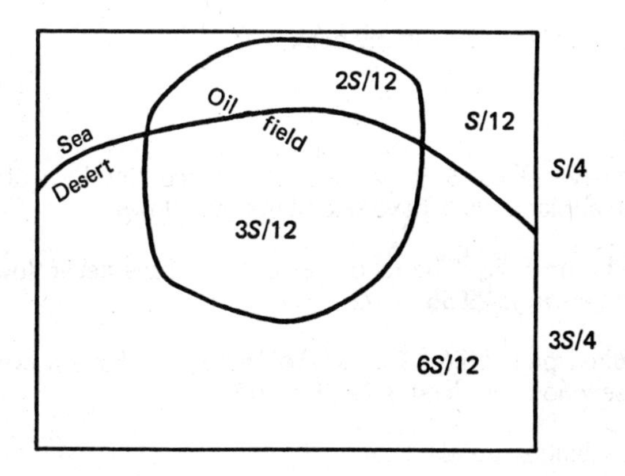

Area under the sea with oil = $(1/4 - x)S$.

Area under the sea without oil = xS.

Area of desert with oil = $3xS$.

Area of desert without oil = $6xS$.

Total area = $S/4 - xS + xS + 3xS + 6xS = S$.

Hence $x = 1/12$.

The required answer is therefore

$$\frac{(1/4 - 1/12)S}{(3/12)S + (1/4 - 1/12)S} = \frac{1/6}{1/4 + 1/6} = 2/5$$

16. Pick a card

Let p be the unknown ratio.

The probability of not drawing a face card in one try is $1 - p$.

The probability of not drawing a face card in three tries is $(1 - p)^3$.

Thus the probability of drawing at least one face card is

$$1 - (1 - p)^3 \quad \text{or} \quad 19/27$$

Hence $(1 - p)^3 = 8/27$, $1 - p = 2/3$, $p = 1/3$.

The proportion of face cards in my deck is 1/3.

17. Poll taking

If 42% have never skied, 58% have done so. Since 29% have already skied *and* taken an airplane, 29% have skied but never flown.

First required probability: The ratio of skiers who have never flown to those who have not flown is 29/58 = 1/2.

Second required probability: The ratio of those who have flown *and* have skied to those who have skied is 29/58 = 1/2.

The two probabilities are identical.

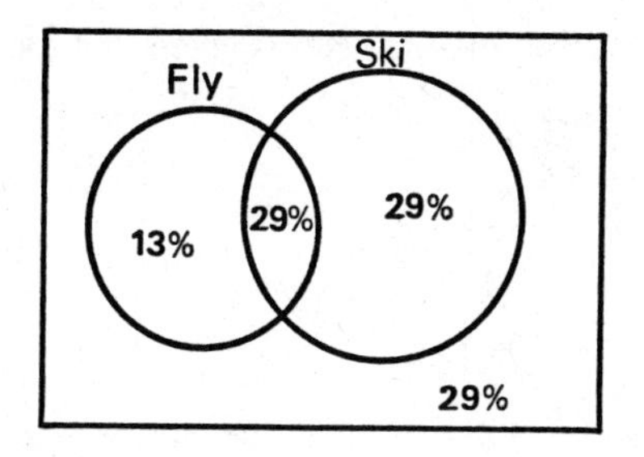

18. Life expectancy

If the one who lives longer dies at 40, both have lived until 40. The corresponding probability is

$$25\% \times 25\% = 1/16$$

If the one who lives longer dies at 50, either the first native lived to 40 and the second to 50, or both lived to 50, or the first lived to 50 and the second to 40. The corresponding probability is

$$(25\% \times 50\%) + (50\% \times 50\%) + (50\% \times 25\%) = 1/2$$

If the one who lives longer dies at 60, neither of the two events whose probability of occurrence we have calculated in fact occurred.

The corresponding probability is

$$1 - 1/16 - 1/2 = 7/16$$

Thus the life expectancy sought is

$$1/16 \times 40 + 1/2 \times 50 + 7/16 \times 60 = 53 \text{ years } 9 \text{ months}$$

19. Red light, green light

Consider two successive lights. There are four possibilities: "both green," "first green, second red," "first red, second green," "both red." Let p_1, p_2, p_3, and p_4 be the four respective probabilities. Since each of these two lights are green two-thirds of the time, we have

$$p_1 + p_2 = 2/3$$

$$p_1 + p_3 = 2/3$$

On the other hand, when the first light is green, the second is green three-fourths of the time:

$$p_1/(p_1 + p_2) = 3/4$$

We know of course that

$$p_1 + p_2 + p_3 + p_4 = 1$$

By solving these four equations with four unknowns, we get

$$p_1 = 1/2 \qquad p_2 = p_3 = p_4 = 1/6$$

The probability of having a red light given that the preceding light was red is therefore

$$p_4/(p_3 + p_4) = (1/6) \div (2/6) = 1/2$$

20. Flight attendants

If we can find at least one dark-haired *and* at least one fair-haired attendant, it is because the three attendants chosen were neither all dark-haired

$$\left(\text{probability } \frac{13 \cdot 12 \cdot 11}{20 \cdot 19 \cdot 18}\right)$$

nor all fair-haired

$$\left(\text{probability } \frac{7 \cdot 6 \cdot 5}{20 \cdot 19 \cdot 18}\right)$$

The required answer is therefore

$$1 - \frac{13 \cdot 12 \cdot 11}{20 \cdot 19 \cdot 18} - \frac{7 \cdot 6 \cdot 5}{20 \cdot 19 \cdot 18} = 0.718$$

21. Sunday paper

Let i be a measure of the annoyance of having two Sunday papers. We are told that the measure of the annoyance of not having a Sunday paper is $2i$. If Mrs. Smith has bought a paper not knowing whether her husband has bought one, she risks incurring a penalty i with a probability of 1/2 (she has no idea what the chances are of her husband getting the paper). This averages out at $i/2$.

If she listens to her son, she will make a mistake one time out of three. She will then incur either a penalty i with a probability $1/3 \times 1/2 = 1/6$ or a penalty $2i$ with the same probability (since the son's guesses are independent of whether the father remembers or forgets). This averages out to $1/6\ (i + 2i) = i/2$.

The degree of annoyance is therefore the same whatever Mrs. Smith does. The two courses of action are equivalent.

22. Raising an army

Since each birth is an independent event and there is equal probability of a boy or a girl being born, the proportion of boys to girls will remain 50 : 50 in spite of the warrior king's proclamation.

23. History test

There are four possible situations:

1. Joe knows the answer and does not lie. What he tells Jim is correct. Probability $3/4 \times 3/4 = 9/16$.
2. Joe know the answer but lies. What he tells Jim is incorrect. Probability $3/4 \times 1/4 = 3/16$.
3. Joe has the wrong answer and does not lie. He gives Jim an incorrect reply. Probability $1/4 \times 3/4 = 3/16$.
4. Joe has the wrong answer and lies. What he tells Jim is correct. Probability $1/4 \times 1/4 = 1/16$.

Jim will therefore get the correct answer 10 times out of 16. He should thus follow Joe's advice.

24. Twins

A pregnant woman can have boy/girl twins only if she has nonidentical twins, which can happen only once in 125 times:

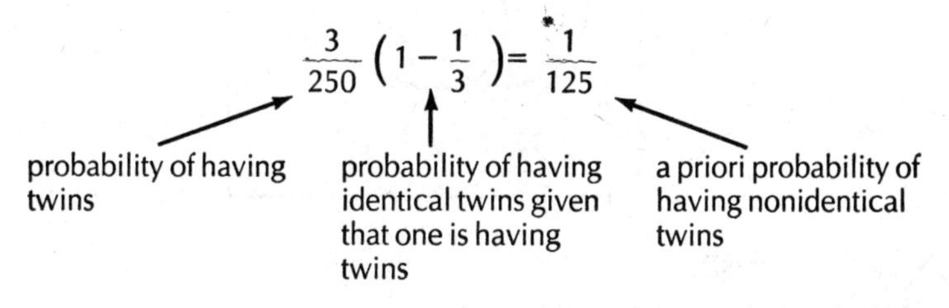

Knowing that she is expecting nonidentical twins, four eventualities are equally possible:

1. The first will be a boy, the second a girl.
2. Both twins will be boys.
3. The first will be a girl, the second a boy.
4. Both twins will be girls.

Two of the four eventualities will produce boy/girl twins. Thus the probability of having boy/girl twins is

$$2/4 \times 1/125 = 1/250$$

25. Jaws

Let x be the percentage of a shark going undetected. We know that the probability of an alarm is 1/30. Hence the probability of having neither an undetected shark nor an alarm is

$$1 - (1/30 + x) = 29/30 - x$$

We have also been told that the probability of false alarm is 10x. Hence the probability of real alarm is 1/30 − 10x. But this is equal to 3x since 3 times as many sharks are detected as go undetected

$$1/30 - 10x = 3x$$

Hence $x = 1/390$

Thus the probability of a peaceful day is

$$29/30 - 1/390 = 0.964$$

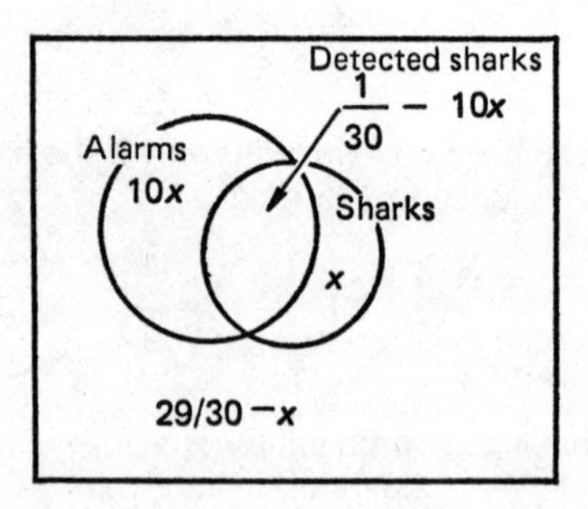

26. Meteorology

There are four possible situations:

1. Fair weather predicted and occurred: probability 1/2 × 2/3 = 1/3.
2. Fair weather predicted but rain occurred: probability 1/5 × 1/3 = 1/15.
3. Rain predicted and occurred: probability 4/5 × 1/3 = 4/15.
4. Rain predicted but fair weather occurred: probability 1/2 × 2/3 = 1/3.

Let i be the measure of the inconvenience for Francine to carry her umbrella around for one whole fine day and $2i$ the measure of inconvenience of not having an umbrella when it rains. If she systematically carries her umbrella every day, the inconvenience will amount to i on two days out of three, hence an average inconvenience of $2i/3$.

If she never takes her umbrella, the inconvenience will amount to $2i$ one day out of three, for an average inconvenience again of $2i/3$.

If she decides to follow the advice of the weather forecasters and takes her umbrella only when rain is predicted, she will suffer an inconvenience of i on 1 day out of 3, and $2i$ on 1 day out of 15, which amounts to an average penalty of $7i/15$ (less than the two previous average penalties). We recommend, therefore, that Francine follow the advice of the weather forecasters.

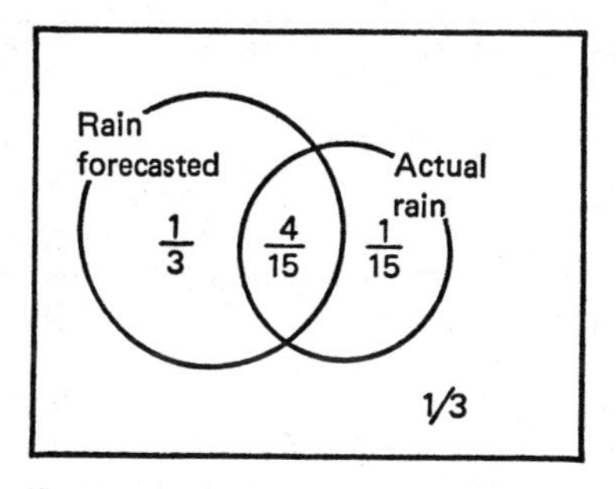

27. Passenger ship

Let us assume that the ship is sailing between the nth and the $(n + 1)$th ports of call. We know that:

- One-fourth of the passengers came aboard at the nth port of call.
- Among this group, one-tenth embarked at the $(n - 1)$th port of call.
- One-fourth of the passengers came aboard at the $(n - 1)$th port of call.

Hence the proportion of passengers who came aboard *either* at the nth *or* at the $(n - 1)$th port of call is

$$1/4 + 1/4 - 1/(4 \cdot 10) = 19/40$$

Hence the proportion of passengers who did not come aboard at either of the two previous ports of call is

$$1 - 19/40 = 21/40 = 52.5\%$$

M_n = berths occupied by the passengers who embarked at the nth port of call

$M_n - 1$ = berths occupied by the passenger who embarked at the $(n - 1)$th port of call

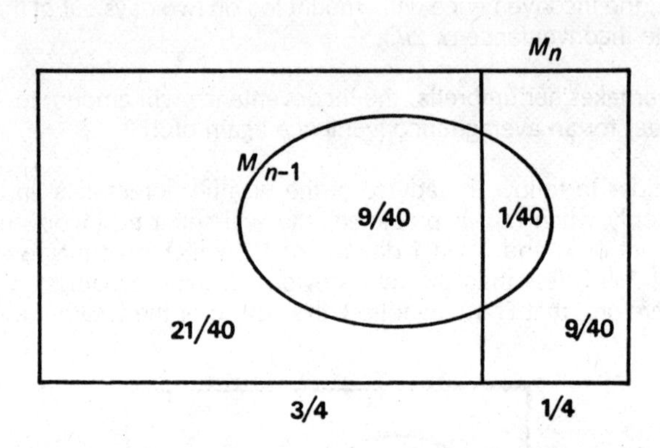

28. Mrs. Martin's umbrella

Since on a fine day Mrs. Martin takes her umbrella one time out of two, and since the probability of a fine day is 0.5, the event "She takes her umbrella on a fine day" will occur with a probability $1/2 \times 1/2 = 1/4$. On the other hand, we know that overall she takes her umbrella on two occasions out of three. The event "She takes her umbrella on a rainy day" will occur, therefore, with a probability $2/3 - 1/4 = 5/12$. But since the probability of a rainy day is 0.5, on 1 occasion out of 12 $(1/2 - 5/12 = 1/12)$, Mrs. Martin will not have her umbrella on a rainy day.

Peter is right.

29. Parking meter

Suppose that I decide not to feed the meter when the café owner thinks that he saw a policeman during the past hour. The café owner makes a mistake one time out of four, in which case the officer may come by during the next 10 minutes with a probability of $10/60 = 1/6$ (since he will inevitably come by in the next hour, not having been by during the preceding hour). The corresponding monetary risk to me is

$$(1/4) \cdot (1/6) \cdot 48 \text{ francs} = 2 \text{ francs, or cost of feeding the meter}$$

The two strategies under consideration are therefore equivalent in cost.

30. Hugh and Caroline

Consider two orthogonal axes. Hugh's time of arrival is plotted on the x-axis, that of Caroline is plotted on the y-axis. The origin corresponds to 11 o'clock for each axis. Each point in the square (0, 0), (0, 1), (1, 1), (1, 0) corresponds to an event with an equal probability of occurrence.

Given Hugh's time of arrival, he will meet Caroline only if she arrived during the previous 15 minutes or will arrive during the next 15 minutes.

The looked-for probability therefore corresponds to the crosshatched area in the figure: that is, to the area of a square of side 1 less that of a square of side 3/4:

$$1 - 9/16 = 7/16$$

Thus there are 7 chances out of 16 that Hugh and Caroline will meet outside the Empire State Building.

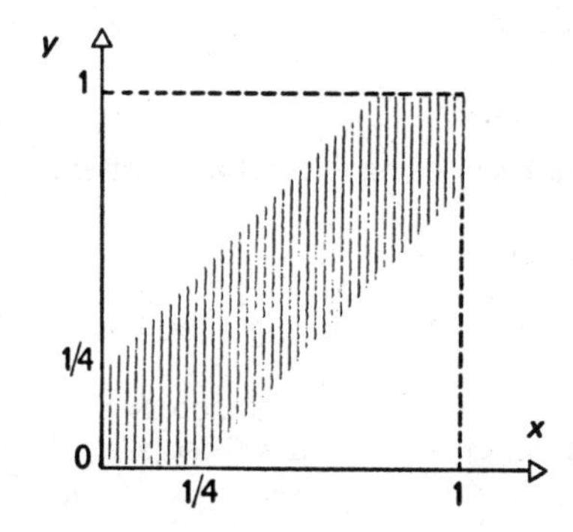

31. Rain or shine?

Consider days $n - 1$ and n.

There are four possible situations.

1. It rains on both days. Probability: $1/4 \times 2/3 = 1/6$.

2. It rains on day n but not on day $n - 1$. Probability: $1/4 - 1/6 = 1/12$.
3. It rains on day $n - 1$ but not on day n. Probability: $1/4 - 1/6 = 1/12$.
4. It does not rain on either of these days. Probability: $1 - 1/6 - 1/12 - 1/12 = 2/3$.

The probability that it does not rain on day n, given that it did not rain on day $n - 1$, is

$$2/3 \div (1 - 1/4) = 8/9$$

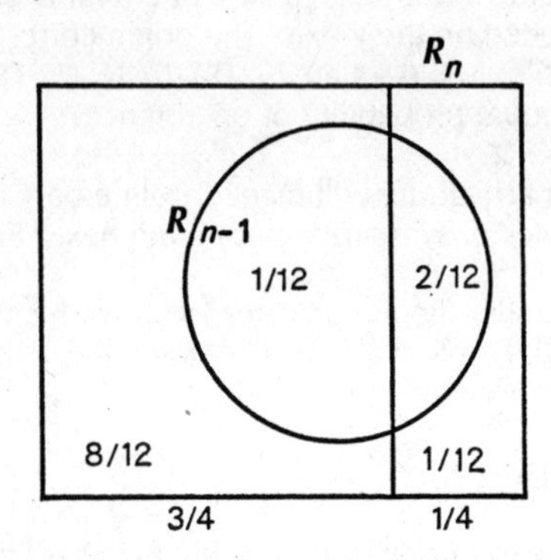

R_n = rain on day n

R_{n-1} = rain on day $n - 1$

32. *Job hunt*

The probability of Mark still being without a job after n letters is

$$(1 - 1/5)^n = (4/5)^n$$

Therefore, the probability of Mark finding a job is $1 - (4/5)^n$.

Mark will stop sending out his résumés as soon as n will be such that this last probability is equal to or greater than 3/4; that is,

$$(4/5)^n \leqslant 1/4$$

Hence

$$n \geqslant \log 4/(\log 5 - \log 4) \qquad \text{or} \qquad n \geqslant 0.602/0.097$$

Hence Mark will have to write seven letters.

33. *Family meal*

On two occasions out of three, the husband will quarrel with his mother-in-law and on one occasion out of three ($1/2 \times 2/3$), he will end up quarreling with his wife.

Similarly, on two occasions out of three, the wife will fight with her mother-in-law and on one occasion out of three she will end up quarreling with her husband.

The proportion of Sundays when the quarrel occurs on both sides (since these quarrels are independent) is

$$1/3 \times 1/3 = 1/9$$

The proportion of Sundays where there is a quarrel between husband and wife is therefore

$$1/3 + 1/3 - 1/9 = 5/9$$

The proportion of Sundays when husband and wife will not quarrel is therefore

$$1 - 5/9 = 4/9$$

34. *Late again?*

Let p_n be the probability that Morris takes the car on his nth trip to the office.

The probability of taking the subway on this nth trip is $(1 - p_n)$.

Morris will take the car on the nth trip if he took it on the $(n - 1)$th trip *and* was not late (one chance out of two), *or* if he took the subway on the $(n - 1)$th trip *and* was late (one chance out of four). Hence we have the following expression for p_n:

$$p_n = (1/2)\, p_{n-1} + (1/4)(1 - p_{n-1}) = 1/4\, (p_{n-1} + 1)$$

Let us express p_n in terms of p_{n-2}, then p_{n-3} down to p_1:

$$\begin{aligned} P_n &= (1/4)(1 + 1/4 + 1/4^2 + 1/4^3 + \ldots + 1/4^{n-2}) + p_1/4^{n-1} \\ &= \frac{(1/4)(1 - 1/4^{n-1})}{1 - 1/4} + p_1/4^{n-1} \\ &= (1/3)(1 - 1/4^{n-1}) + p_1/4^{n-1} \end{aligned}$$

When n becomes large, p_n tends toward 1/3.

Morris therefore has one chance out of three of arriving late at his office on his 467th trip.

35. The dailies

Let x be the unknown percentage.

Let V be the ensemble of all the adults on vacation:

> S the subset of readers of the *Sun*
> P the subset of readers of the *Post*
> T the subset of readers of the *Times*

We can draw the corresponding Venn diagram and write in the percentages of vacationing adults for each zone.

The entire V ensemble obviously has a value of 100%; hence

$$42 + (14 + x) + (12 + x) + (15 + x) + (11 - x)$$
$$+ (2 - x) + (3 - x) + x = 100$$

Hence $x = 1\%$. One percent of the vacationing adults read all three dailies.

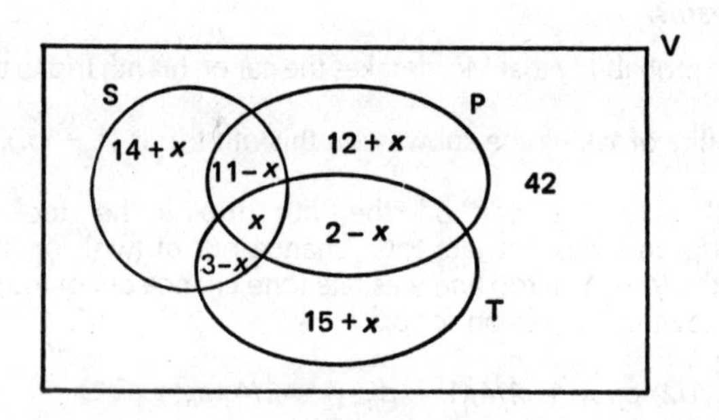

36. International gathering

Number of ways to seat the French: 4!.

Number of ways to seat the Africans: 2!.

Number of ways to seat the Russians: 3!.

Number of ways to seat the Swedes: 4!.

Hence the number of ways to seat these 13 people without altering the relative positions of the national groupings is $4! \cdot 2! \cdot 3! \cdot 4! = 6912$.

The number of ways to seat the groups on a bench is 4!.

The total number of possible seating arrangements on a bench is therefore

$$4! \times 6912 = 165,888$$

The number of ways to seat the groups around a round table is 3!.

The total number of possible seating arrangements around a round table is therefore

$$3! \times 6912 = 41,472$$

Note: If F is the French group, A the African, R the Russian, and S the Swedish, the possible seating arrangements around a round table are

FSAR, FARS, FASR, FRAS, FRSA, FSRA: total 6 or 3!

On a bench, there are four times as many possible seating arrangements because we have, for instance, $RAFS \neq FSRA$, which is not true in the case of the round table.

37. Sensible or sensitive?

Let i be a measure of the inconvenience of giving a wrong answer. By not copying what her neighbor has written, the foreign student stands to lose i on one occasion out of two, that is, on the average $i/2$. By copying what her neighbor has written, she stands to lose $3i$ 1 time out of 10 (when she is caught red-handed). She also stands to lose i when her neighbor has the wrong answer *and* she is not caught red-handed. This will happen with a probability

$$1/5 \times 9/10 = 9/50$$

The average risk is therefore

$$3i/10 + 9i/50 = 24i/50$$

This is slightly less than the previous risk ($i/2 = 25i/50$).

By considering the average risk, the foreign student will see that there is an advantage to be gained by copying from her neighbor.

38. Vacation memories

Let x be the number of photographs of castles.

Let y be the number of photographs of triumphal arches.

Let z be the number of photographs of waterfalls.

Let t be the number of photographs of medieval churches.

The sights that I did *not* photograph include: 0 triumphal arches, $(3y - t)$ medieval churches, x castles, and $(2x - z)$ waterfalls.

We know that one-fourth of the photos depict castles:

$$x = 1/4\ (t + x + y + z)$$

and that half the sights have been photographed:

$$t + x + y + z = (1/2)(3y + 2x + 2x + y) = 2x + 2y$$

Hence $t + z = x + y$.

Substituting in the first equation we get

$$x = y$$

The proportion of photos of triumphal arches in my collection is therefore

$$y/(x + y + z + t) = 1/4$$

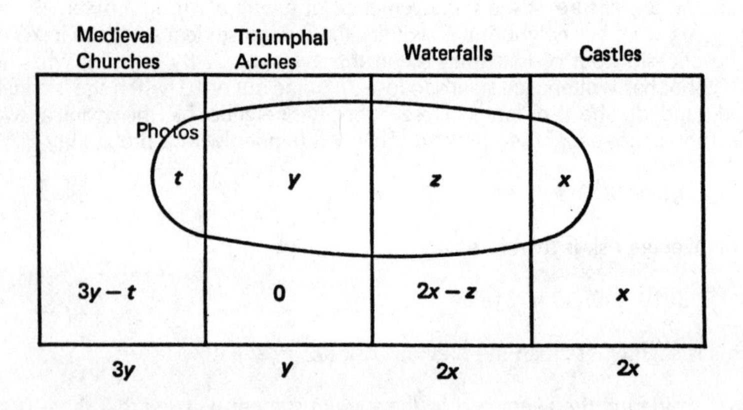

39. Washing-up time

Number of ways of choosing five people among 20:

$$C_{20}^{5} = \frac{20!}{5!\,15!}$$

Number of ways of choosing these five people without picking married couples:

$$C_{10}^{5} \times 2^5 = 2^5 \cdot \frac{10!}{5!\,5!}$$

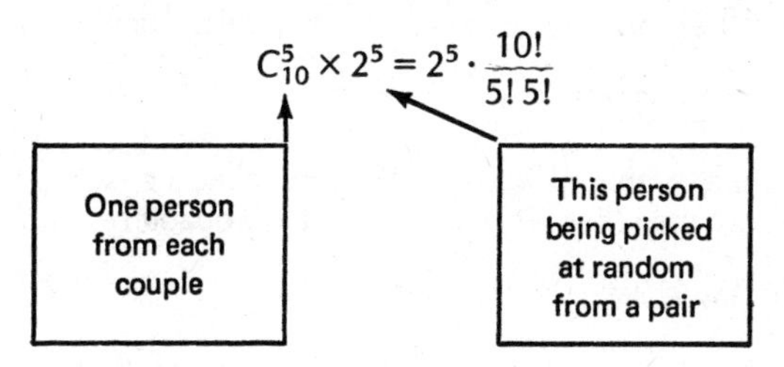

Required probability:

$$\frac{C_{10}^{5} \cdot 2^5}{C_{20}^{5}} = \frac{6 \cdot 7 \cdot 8 \cdot 9 \cdot 10}{16 \cdot 17 \cdot 18 \cdot 19 \cdot 20}\, 2^5 = \frac{168}{323} = 0.52$$

40. Homework

Since 1 pupil out of 3 is quizzed each time and since 1 pupil out of 10 of those quizzed was quizzed the last time, 1 out of 30 will be quizzed on successive occasions. Thus there is a 1/30 risk of being scolded by Madame Dubois. A pupil will be quizzed at least once during one of the two classes with a probability of

$$1/3 + 1/3 - 1/30 = 19/30$$

On 11 out of 30 occasions, he will not be quizzed during either class, since $1 - 19/30 = 11/30$, and he will have learned his lesson for nothing. The chances that the lazy pupil will learn a lesson for nothing are 11 times greater than the chances of his being scolded.

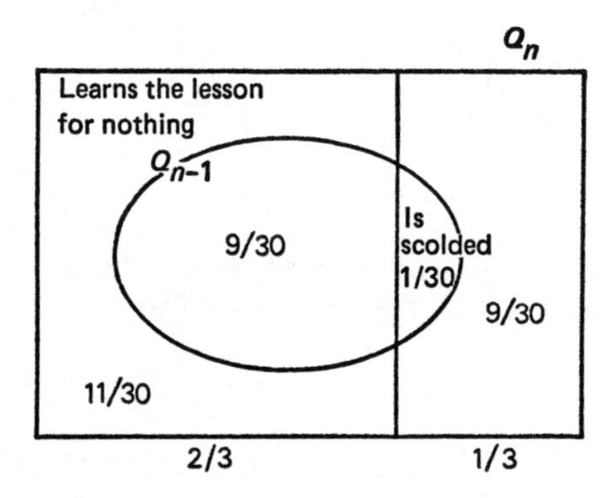

Q_n = quizzed during class n

Q_{n-1} = quizzed during class $n - 1$

41. *Twin sisters*

If Ann is on the left when both girls claim to be Bridget, the twin on the left is fibbing and the other is telling the truth. This will occur with a probability

$$1/5 \times 3/4 = 3/20$$

But the fact that both girls claim to be Bridget means that one of the two is fibbing and not the other. This occurs with a probability

$$1/5 \times 3/4 + 4/5 \times 1/4 = 7/20$$

The required probability is the conditional probability that the event "Ann the twin to the left" occurs, given that both girls claim to be Bridget, which will be

$$(3/20) \div (7/20) = 3/7$$

42. *At Le Mans*

The probability that Ford will lose:

$$2/(2 + 2) = 1/2$$

Hence the probability that Ford will win: 1/2.

The probability that Jaguar will lose:

$$5/(5 + 1) = 5/6$$

Hence the probability that Jaguar will win:

$$1 - 5/6 = 1/6$$

Therefore, the probability that either Ford or Jaguar wins is

$$1/2 + 1/6 = 4/6 = 2/3$$

This is the same as the probability that Maserati will lose. Thus there are two chances to one that Maserati will not come in first.

Ford wins	Jaguar wins	Maserati wins
2/(2+2)	1/(5+1)	1/(2+1)

2. Logically speaking

1. Differential voting methods

An electoral district has 100,000 voters. Three candidates are running for office: *A, B,* and *C.*

The local voters have divided opinions over which candidate is best:

1. A first group, comprising 33,000 voters, favors candidate *A*, followed by *B*, then *C*.
2. A second group, comprising 18,000 voters, favors candidate *B*, followed by *A*, then *C*.
3. A third group, comprising 12,000 voters, favors candidate *B*, followed by *C*, then *A*.
4. A fourth group, comprising 37,000 voters, favors candidate *C*, followed by *B*, then *A*.

Question: Who will be elected under each of the following rules?

1. Whoever gets the most votes is elected.
2. After a first go-round, where voting takes place in the usual way, there is a runoff between the two candidates with the greatest

number of votes, all voters in the second round being allowed to vote for one of these two candidates only.

3. There are three go-rounds: in the first, *A* and *B* are the only candidates; in the second, *B* and *C* are the only candidates; in the third, *C* and *A* are the only candidates. This procedure gives the general preferances for the district and in particular selects the best candidate.

2. Town council

A town council convenes to determine the best system for sprinkling the local streets. Two methods are proposed: *A* and *B*. Each council member belongs to one of four parties: Independent, Democratic, Republican, and Socialist, and each political party is represented by an equal number of council members. The mayor observes the following:

1. All the Independent members agree on the same method while opinions are mixed in the other three groups.
2. As many Democrats favor *A* as Republicans favor *B*.
3. A third of the council members who are partial to method *B* are Socialists.

What type of sprinkling system will the mayor order if he abides by the majority voice?

3. Resort story

Five vacationers meet at a resort.

Emilie says: "I live in Acapulco. So does Ben. Peter lives in Pittsburgh."

Ben says: "I live in Boston. So does Charles. Peter lives in Pittsburgh."

Peter says: "I don't live in Pennsylvania. Neither does Emilie. And Melanie lives in Miami."

Melanie says: "My father lives in Acapulco, my mother lives in Pittsburgh, and I live in Chicago."

Charles says: "Emilie comes from Acapulco. So does Ben. I live in Chicago."

During this conversation all of the five vacationers tell two truths and one lie. Where, in fact, does each of them live?

4. Stamps and sharks

A group of 17 people have the following characteristics:

1. All those who collect stamps and were born in Hawaii like to go shark hunting.
2. No one weighing over 100 kg collects stamps.
3. Everyone born in Hawaii likes to hunt sharks or collect stamps.
4. No one weighing under 100 kg and not collecting stamps was born in Hawaii.

How many group members were born in Hawaii but don't care for shark hunts?

5. Blue jeans

Ann, Beatrice, and Charlotte notice one day that they are wearing identical jeans. How are they dressed, assuming that Ann owns a pair of straight-leg jeans with pockets and a faded pocketless pair, Beatrice owns a pair of pocketless jeans and a pair of faded straight-legs with pockets, and Charlotte owns a pair of flared jeans and a pair of dark blue straight-legs with pockets?

6. Black hat, white hat

"Look carefully," Mr. Perkins tells his three students. "Here are five hats: three white and two black. Now close your eyes and I'll give you each a hat. When you open your eyes again, you'll each be able to see what the other two are wearing but not what's on your own head. Nor will you see the hats which haven't been used. The first one to figure out the color of his hat will get a dollar." After a few moments, all three students independently arrive at the conclusion that their hats are white. Why do they think so?

7. Des jeunes gens charmants (some charming young people)

Although I was a brilliant French student, I must admit that I have trouble understanding the French when they speak. One day, on a motor trip through France, I picked up a group of hitchhikers who couldn't speak English. Everything they said seemed to me to have two possible meanings (the second is noted here in parentheses). The first said: "We're on our way to Spain (we come from Nancy)." The second said: "We're not going to Spain and we come from Nancy (we stopped over in Paris but we are not going to Spain)." The third said: "We aren't from Nancy (we stopped over in Paris)." *Qu'est-ce-que ces jeunes gens charmants ont vraiment fait?*

8. In cannibal country

Three young couples, tired of their usual vacations at their parents' country houses, decide to go into the African bush. Unfortunately, they are captured by cannibals, who weigh them before preparing them for eating. The total weight of the six tourists is not a whole number but the wives together weigh exactly 171 kg. Leon and his wife weigh the same, Victor weighs one and a half times more than his spouse, and Maurice weighs twice what his wife does. Georgette weighs 10 kg more than Simone, who weighs 5 kg less than Elizabeth. Fortunately, during the long weighing process, five of the six captives escape. But Elizabeth's husband is eaten. How much did he weigh?

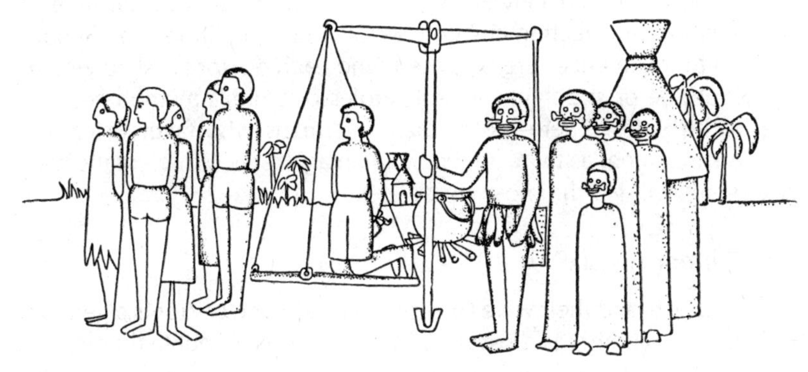

9. Target practice

Three friends decide to do some target shooting.

"I'll bet," says the first, "that at least one of you two won't hit the bull's-eye on your first shot."

"And I'll bet," says the second to the first, "that if you hit the bull's-eye on your first shot, you'll win your bet."

"As for me," says the third, "I bet that all three of us will hit the bull's-eye the first time around."

They fire. Is it possible that the second and third gunmen can both either win or lose their bets on the first shot?

10. Spanish-speaking mothers

"Every mother of seven who speaks Spanish has a chignon even if she wears glasses." "Every woman wearing glasses has seven children or can speak Spanish." "No woman who does not have seven children wears glasses unless she has a chignon." "Every woman with seven children and glasses speaks Spanish." "No woman with a chignon has seven children." If all five of these statements are true, what does it tell you about those women who wear glasses?

11. Doctors' convention

An international convention of dermatologists includes specialists from Great Britain, Germany, and France. There are twice as many Frenchmen as Germans, and there are twice as many Germans as Englishmen. Two methods for treating Gilbert's serious skin condition—the Simon and Smith treatments—are proposed, and each doctor is asked which procedure he or she thinks is best. Professor Smith's method gets all the British votes. There are as many Germans who favor Professor Simon's treatment as there are Frenchman against it. Which of the two treatments will get the most votes?

12. Dinner for six

Three doctors and their wives meet for dinner. One doctor is a general practitioner, another is a psychiatrist, the third is an ophthalmologist.

Mrs. Brown, who is quite tall, exceeds the general practitioner in height by as much as he exceeds his wife in stature. The wife whose height is the closest to the general practitioner's weighs the same as he does, while Mrs. Burns weighs 10 kg less. Dr. Black is 20 kg heavier than the ophthalmologist. What is the psychiatrist's name?

13. Bicycle race

A bicycle race attracts as many Italian racers as Swedes. The first group is generally dark-haired and small, the second is generally blond and tall. However, one-fifth of the Swedes are small and dark. We also know that overall there are three small dark-haired bikers for each two tall blonds and that none of the cyclists are tall and dark-haired or small and blond.

Two spectators wait for the first arrivals at the finish line. A blond and a tall racer arrive.

"Here comes a Swede," says *A*.

"Not necessarily," says *B*.

If you were there would you be as sure of yourself as *A*? Or would you share *B*'s doubts?

14. First cousins

Three brothers, Peter, Paul, and Jack, meet with their children in the family country house. The children announce the following in turn:

Isabelle: "I'm three years older than John."

Theresa: "My father's name is Jack."

Ivan: "I'm two years older than Isabelle."

Marie: "I'd rather play with one of my cousins than with my brother."

Catherine: "My father's name is Peter."

Ann: "I get along best with Uncle Jack's son."

John: "My father and his brothers each have fewer than four children."

Frank: "My father is the one with the least children."

Can you deduce the first name of each of the cousins' fathers given the facts above?

15. Nighttime bottle

One Saturday evening a young couple decide to go to the movies, leaving their two children, a 3-year-old son and a 6-month-old daughter, with their grandmother. The grandmother wonders if the baby has been given her nighttime bottle but has no way of knowing a priori. She could ask the older child but there is one chance out of four that the boy would answer incorrectly. Knowing that it is twice as inconvenient for a 6-month-old to miss a nightly bottle than to receive two, what would you do if you were the grandmother?

1. Give the baby a bottle without asking the 3-year-old?
2. Question the 3-year-old and only give the nighttime bottle if he says that his sister has not yet received it?

16. Election talk

Three brothers discuss the coming local election.

Peter announces: "If John votes for Smith, I'll vote for Jones. But if he votes for Brown, I'll vote for Smith. On the other hand, if Jack votes for Jones, I'll vote for Brown."

John replies: "If Peter votes for Brown, I won't vote for Jones. But if Jack votes for Smith, I will."

Jack announces: "If Peter votes for Jones, I won't vote for Brown."

On election day they each vote for a different candidate. Who votes for whom?

17. Journey to Mars

Advanced technology proves that Mars is uninhabited except for two large cities, Mars-Town, whose citizens never lie, and Mars-City, whose inhabitants never tell the truth. Since Martians can circulate freely, the citizens of one of these two urban centers can often be found in the other. Two cosmonauts touch down in one of the cities but do not know which it is. The first cosmonaut hails a native who is passing by the spacecraft and asks in Martian if they are in Mars-City.

"No," the Martian replies. (He knows how to answer only yes or no and could be lying.)

The second cosmonaut then asks a very clever question, the answer to which will prove where they are. What is his question?

18. Hearing

Three robbery suspects are questioned before a judge. One (the thief) consistently lies; another (his accomplice) conceals only part of the truth; the third (who has been falsely accused) always tells the truth.

During the hearing, the judge tries to establish the suspects' professions. Bernard says: "I'm a house painter. Alfred tunes pianos. Charles is a decorator."

Alfred says: "I'm a doctor. Charles sells insurance. Bernard will undoubtedly tell you that he paints houses."

Charles says: "Alfred tunes pianos. Bernard is a decorator. I sell insurance."

Based on the information above, the presiding judge tries to establish what the accomplice does for a living. Can you help her?

19. Stickup

After a stickup, a gangster flees across three rivers, each of which can be crossed by bridge. At the end of each bridge lie three roads, one to the right, one to the left, and one leading straight ahead. The police manage to catch the gangster's accomplice and ask where the thief has gone. They receive the following answer: "After the first bridge he turned to the right. After the second he didn't turn right. After the third he didn't turn left." Knowing that two of the accomplice's statements are false and that the gangster took each of the three possible directions once, what was his route?

20. Faulty reporting

Three journalists watch a celebrity dine. Afterward they file the following stories:

Jules: "He had a cocktail, then ordered roast turkey and dessert. He finished with a cup of coffee."

Jack: "He skipped cocktails, ordered steak, and ended with a soufflé."

Jim: "He had a cocktail, ate a steak, had strawberry sherbert for dessert, then ordered coffee."

One of the reporters filed nothing but false information, another reported only one item incorrectly, the third gave a fully accurate account. Can you deduce from the three conflicting reports what the celebrity ate for dessert?

21. Amateur doctor

Aspirin does wonders for my headache and helps my rheumatic knee but it gives me nausea and upsets my stomach. Herbal medicine cures my nausea and stomach upset but gives me hip pain. Antibiotics sooth my headache and nausea but irritate my stomach and knee and give me a stiff neck. Cortisone helps my stiff neck and rheumatic knee but aggravates my hip condition. Hot compresses work wonders for my upset stomach and stiff neck. I woke up today with a pounding headache which prevents me from figuring out how best to doctor myself. What do you think I should do?

22. Fibs

John and Peter occasionally lie.

John tells Peter: "When I don't lie, you don't either." Peter replies: "When I lie, so do you."

During the foregoing exchange, can one of them be fibbing but not the other?

23. April Fool!

Three teachers sit chatting on a bench during a recess break. They are unaware that their students have pinned April Fool signs to their backs. When they stand up to return to class, all three teachers burst out laughing. Each is sure that the other two are laughing at each other and that he is not part of the joke. Suddenly, one of the three stops laughing. He has just realized that he has a sign on his back, too. How did he figure that out?

24. Terrorist

On Christmas Day, 1988, on January 1, 1989, and on the following July 4, the president of the United States was the object of a series of attempted assassinations, each of which he escaped by a miracle. One attempt was carried out by Ruritanians, one by Bothians, and one by Trinians.

The police did not know exactly which group was responsible for any given attack on the president's life. An international terrorist who was arrested and interrogated at about that time made the following declaration:

"The Christmas attempt was made by the Trinians, but not the attack of January 1. The July 4 attempt cannot be attributed to the Bothians." Knowing that of the three statements made by the terrorist, one was true and two were false, can you help the police in identifying the group responsible for each attempted murder?

25. Conundrum

Lewis Carroll liked to pose the following four-part problem:

"Either the criminal arrived by car or the witness was mistaken. If the criminal had an accomplice, he arrived by car. Either the criminal did not have an accomplice or a key, or he had an accomplice and a key. The criminal had a key." What can one deduce from the foregoing?

26. Three kings from Orient

On the way to see Baby Jesus, Melchior, Gaspar, and Balthazar stop by the roadside. While they are resting, they discuss the order in which they will arrive at the manger.

Melchior: "If I'm last, Gaspar won't be first, and if I'm first, Gaspar won't be last."

Balthazar: "If I'm last, Melchior won't be after Gaspar, and if I'm first, Melchior won't be before Gaspar."

Gaspar: "If I'm neither first nor last, Melchior won't be before Balthazar."

The Holy Ghost managed to find a way to respect each king's conclusion. In what order did the three men arrive?

27. Payday

Three executives discuss their monthly salaries.

Brown: "I earn $6000 a month, which is $2000 less than Smith and $1000 more than Jones."

Smith: "I'm not the lowest paid. The difference between Jones' salary and mine is $3000. And Jones earns $9000 a month."

Jones: "I earn less than Brown, who earns $7000 a month. Smith earns $3000 more than Brown."

Each of the executives has told two truths and one lie. Can you tell me what Brown, Smith, and Jones earn per month?

28. High jumps

Three high school students are practicing high jumps. The bar has just been raised to 1.3 meters.

The first girl says to the second: "I bet that I'll only clear the bar if you don't." The second student bets the same with the third, who in turn bets the same with the first. Is it possible that none of the three girls will lose her bet?

29. Flowered hats

If every Englishwoman wears a flowered hat when she visits the Empire State Building and every woman in a flowered hat who visits the Empire State Building is from England, does this mean that every Englishwoman owning a flowered hat visits the Empire State Building?

30. Aunt Ernestine

Aunt Ernestine is coming for Sunday lunch. My wife has gone out in her car. Will she remember to pick up Aunt Ernestine, who cannot go out alone, after she finishes her shopping? I have no way of knowing this a priori. Perhaps I should take my car and drive the 10 km to Aunt Ernestine's to fetch her. But that might not be necessary. I could call Aunt Ernestine, of course, and fetch her only if she answers the phone. But since she is extremely deaf, I've noticed that two times out of five she doesn't answer the phone when it rings. It's three times more vexing to forget Aunt Ernestine than to go and fetch her unnecessarily. I have three possible choices:

Choice 1: stay home
Choice 2: go to Aunt Ernestine's regardless
Choice 3: go and fetch her only if she answers the phone

What would you do in my place?

31. Apartment complex

Twelve rectangular apartment blocks have just been built outside our Western town. Each block is bordered by roads. A mailbox stands at every corner and every crossroad. The apartment blocks are connected either at their corners or along an entire side. Knowing that there are a total of 37 road segments, and hence 37 spaces between mailboxes, exactly how many mailboxes does the entire complex contain?

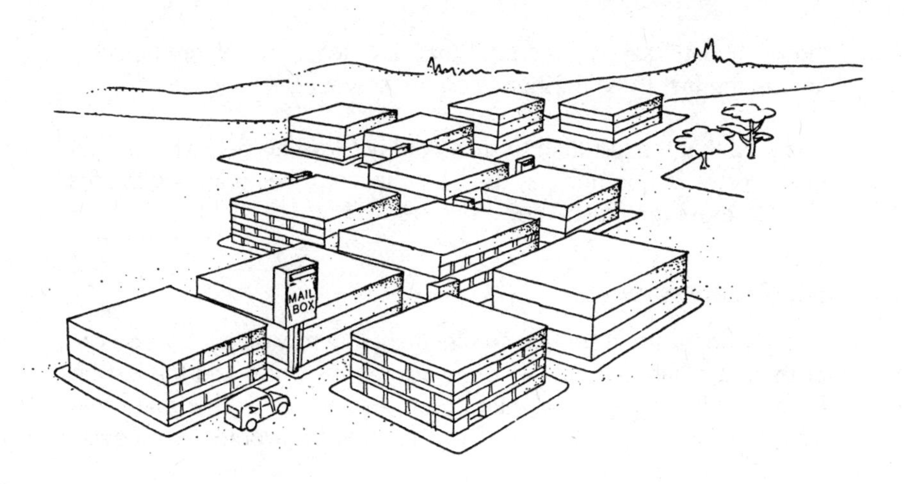

32. French holiday

Aunt Julia takes her three nephews to Paris. When the boys return, they tell their friends the following:

Nephew One: "We went up the Eiffel Tower but didn't visit Versailles. We also saw the Arch of Triumph."

Nephew Two: "We went up the Eiffel Tower and visited Versailles but we didn't see the Arch of Triumph or go to the Louvre."

Nephew Three: "We didn't go up the Eiffel Tower but we saw the Arch of Triumph."

Knowing that each boy told one lie, what did they actually see on their trip?

33. Horse race

Two friends discuss their bets before a horse race. One has bet on Xavier and Yellow Ribbon, the other on Zephyr.

"I'll bet," says the latter, "that if my horse is among the first three to finish, Xavier will be, too."

"And I'll bet," says his friend, "that if at least one of my horses is among the first three finishers, you'll lose your bet."

Knowing that the last man to speak did not lose his bet, which of the three horses in question has the best chance of being among the first three to cross the finish line?

34. Prisoners' wives

Five prisoners are in solitary confinement in a row of five towers. Two of them are habitual liars, two always tell the truth, and the last lies occasionally. They like to talk about their wives to the guard who brings them bread and water. They describe the women as follows:

Inmate One: "My beloved Ann is blond. So is the wife of the man in the next tower."

Inmate Two: "My dear wife Beatrice is a redhead. The wives of the prisoners on either side of me are strawberry blonds."

Inmate Three: "My sweet Claire has red hair. So do the wives of the men on both sides of me."

Inmate Four: "My precious Gertrude is redheaded. The wives of my two neighbors are strawberry blonds."

Inmate Five: "My beautiful Joanna is brunette. So is the wife of the man next door. The wife of the man in the tower at the other end is neither brunette nor redheaded."

What colors are Ann, Beatrice, Claire, Gertrude, and Joanna's hair?

Logically yours

1. Different voting methods

If one chooses the first method with only one go-round, *A* will get 33,000 votes, *B* 30,000, and *C* 37,000. *C* is therefore elected.

Second method: C and A are confronted in a second go-round. A obtains 51,000 votes (33,000 + 18,000) and C obtains 49,000 votes (12,000 + 37,000). A is therefore elected.

Third method:

A and B confrontation: The latter wins, 67,000 votes versus 33,000.

B and C confrontation: B wins, 63,000 votes versus 37,000.

C and A confrontation: A wins, 51,000 votes versus 40,000.

For the whole district, B is therefore considered the best candidate, followed by A, followed by C; that is, B is elected.

Thus, without any change in public opinion, the voting method can determine which candidate will be elected.

We are quoting here from a course on game theory by Professor Jean Bouzitat.

2. *Town council*

Let x be the number of town council members belonging to each political party and y be the number of Democrats who are in favor of method B (which is also the number of Republicans in favor of method A).

First hypothesis: All the Independents are in favor of method B. Votes for B: x Independents; y Democrats; $x - y$ Republicans; $1/2\ [x + y + (x - y)]$ Socialists or x Socialists.

This implies, however, that all Socialists must be of the same opinion, which is impossible.

Second hypothesis: All the Independents are in favor of method A. Votes for B: 0 Independents; y Democrats; $x - y$ Republicans; $1/2\ [y + (x - y)]$ Socialists or $x/2$ Socialists.

As a result, the number of votes for method B is $3x/2$ and the number of votes for method A is

$$4x - 3x/2 = 5x/2$$

The mayor will order the adoption of method A. (See figure, page 50.)

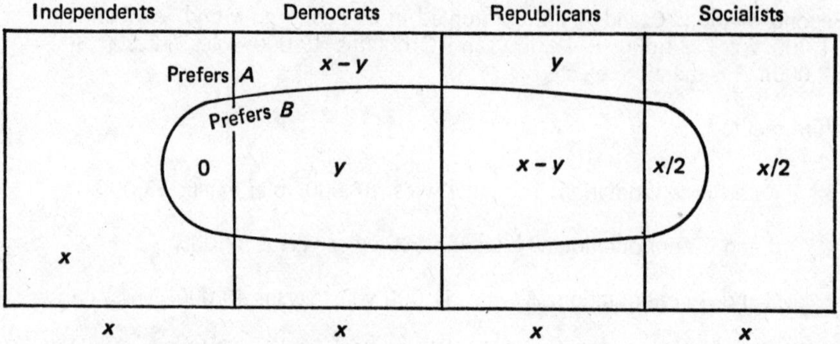

Independents Democrats Republicans Socialists

Prefers A $x-y$ y

Prefers B

0   y $x-y$ $x/2$ $x/2$

x

x x x x

3. *Resort story*

1. Let us suppose that Peter does not live in Pittsburgh. Emilie and Ben must be telling the truth in their other two statements. Therefore, Ben lives both in Acapulco and in Boston, which is impossible. Hence the original supposition was false and Peter does not live in Pittsburgh. Hence regarding what Peter says, Melanie lives in Miami.

2. Let us suppose that Emilie does not live in Acapulco. She is therefore lying when she speaks of herself and her other two statements are true. Therefore, Ben lives in Acapulco. Hence, Ben is lying when he states that he lives in Boston. This means that Charles lives in Boston. Charles therefore lies twice (first and third statements), which is impossible. Our original supposition was false and Emilie lives in Acapulco.

Since Emilie only lies once, Ben does not live in Acapulco. Charles, who also can lie only once, lives in Chicago, and Ben, for the same reason, lives in Boston.

4. *Stamps and sharks*

Consider the set E of all those who were born in Hawaii and don't like to hunt sharks. Let us divide the sample into two groups:

> *A:* Subset of the stamp collectors.
> *B:* Subset of the non-stamp collectors.

The first pronouncement tells us that all those who collect stamps and who are born in Hawaii like to hunt sharks. The subset A is therefore empty.

The third pronouncement tells us that all those who are born in Hawaii like to hunt sharks or collect stamps. Subset B is therefore also empty.

The set E consisting of A and B is therefore also empty. The required answer is 0.

5. Blue jeans

Consider the three following subsets of blue jeans:

$F =$ faded blue jeans
$S =$ straight-leg jeans
$P =$ jeans with pockets

Note that F, S, and P are complementary subsets.

There will thus be $2^3 = 8$ types of jeans possible. We must determine which of these types match the jeans owned by Ann (A), Beatrice (B), and Charlotte (C). We can draw up the following matrix:

FSP	$FS\bar{P}$	$F\bar{S}P$	$F\overline{SP}$	$\bar{F}SP$	$\bar{F}S\bar{P}$	$\bar{F}\bar{S}P$	$\overline{FSP}$
A	A		A	A			
B	B		B		B		B
		C	C	C		C	C

The only blue jeans whose characteristics are common to those owned by the three girls, A, B, and C, correspond to column four: that is, faded, flared, and pocketless.

6. Black hat, white hat

Let S_1, S_2, and S_3 be Mr. Perkins's three students. The first, S_1, reasons along the following lines:

"My two friends have white hats. Mine is either white or black. If it was black, S_2 would say 'S_1 has a black hat, S_3 has a white hat. I must have a white hat because if it were black, S_3, seeing only two black hats, would announce immediately that he had a white hat. But he has not done this. He is hesitating about the color of his hat. Therefore, my hat is white.' But S_2 says nothing. This must therefore be because my own hat cannot be black. It must be white."

Each of the three students could have reasoned the same way, assuming that the other two possessed ample capabilities for logical thought.

7. Des jeunes gens charmants (some charming young people)

We note first, from what the second hitchhiker said, that the trio are not going to Spain (whichever statement is the correct one). The second statement of the first hitchhiker was therefore the correct interpretation: they were coming from Nancy. It is therefore necessary to consider the second statement of the third hitchhiker as the one that is accurate: The trio stopped in Paris.

To summarize: The three hitchhikers came from Nancy, stopped in Paris, but were not going to Spain.

8. *In cannibal country*

Consider first the three wives (total weight 171 kg). Let s be Simone's weight. Elizabeth weighs $s + 5$ and Georgette weighs $s + 10$. We therefore have the equation

$$3s + 15 = 171$$

Hence $s = 52$. Elizabeth weighs 57 kg and Georgette weighs 62 kg. Note that Elizabeth's weight is the only weight represented by an odd number.

Consider next the three husbands. The total weight of the three couples is not a whole number. However, we know that one of the husbands weighs the same as his wife, another (Victor) is one and a half times his wife's weight, and the third husband weighs twice as much as his wife. Victor must therefore be the husband of a woman whose weight, expressed in kilograms, is an odd number. Therefore, he must be Elizabeth's husband. It is he who will be eaten. He weighs 85.5 kg.

9. *Target practice*

Let C_1 = statement: the first friend scores a bull's-eye on his first shot.

Let C_2 = statement: the second friend scores a bull's-eye on his first shot.

Let C_3 = statement: the third friend scores a bull's-eye on his first shot.

The second friend says that if C_1 is true, it is not possible for C_2 and C_3 to be true simultaneously. He will lose his bet if, and only if, all three statements are true, that is, if the third friend wins his bet. It is therefore not possible for the second and third friends to win both or lose both their bets simultaneously.

10. *Spanish-speaking mothers*

After the fifth sentence, we know that if a woman is the mother of seven children, she does not have a chignon. The first sentence tells us that any Spanish-speaking mother with seven children has a chignon. Therefore, no mother with seven children speaks Spanish. But the fourth sentence states that any woman with seven children who wears glasses speaks Spanish. Hence there are no women with seven children among those who wear glasses. We can therefore deduce from the second sentence that all those who wear glasses speak Spanish and from the third sentence that they also have chignons.

To conclude, if the five statements are accurate, all women with glasses speak Spanish and have a chignon but none of them have seven children.

11. Doctors' convention

Let x be the number of British specialists, $2x$ the number of German specialists, and $4x$ the number of French specialists.

Let y be the number of German specialists in favor of the Simon treatment. The number of votes for Smith are:

$$\underbrace{x}_{\substack{\text{number}\\ \text{of}\\ \text{British}}} + \underbrace{(2x - y)}_{\substack{\text{number}\\ \text{of}\\ \text{Germans}}} + \underbrace{y}_{\substack{\text{number}\\ \text{of}\\ \text{French}}} = 3x$$

Number of votes for Simon:

$$\underbrace{y}_{\substack{\text{number}\\ \text{of}\\ \text{Germans}}} + \underbrace{(4x - y)}_{\substack{\text{number}\\ \text{of}\\ \text{French}}} = 4x$$

The Simon treatment will thus get more votes than the Smith treatment.

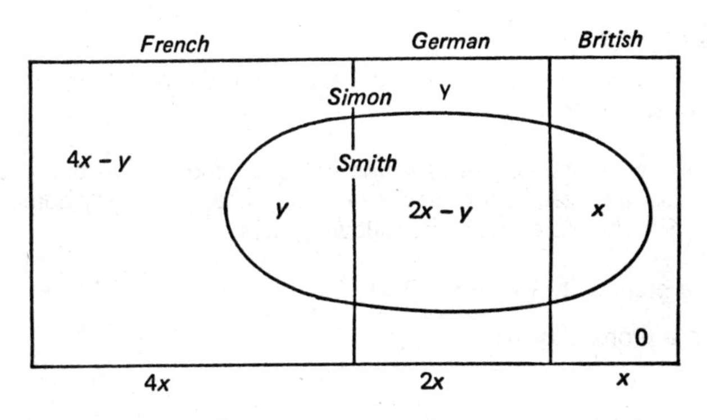

12. Dinner for six

Mrs. Brown and the general practitioner's wife have heights that are equidistant from the height of the general practitioner. Therefore, the wife whose height is nearest to that of the latter is neither Mrs. Brown nor Mrs. Burns (who weighs 10 kg less). It must therefore be Mrs. Black. The wife of the general practitioner can therefore be none other than Mrs. Burns and the general practitioner must be Dr. Burns. Dr. Black cannot be the ophthalmologist (since he weighs 20 kg more). He must therefore be the psychiatrist.

13. Bicycle race

Since one-fifth of the Swedish cyclists are small and dark, four-fifths must be tall blonds. Two runners out of five are therefore simultaneously tall, blond, and Swedish.

On the other hand, we know that among all the cyclists there are three small, dark-haired contestants for every two tall blonds. This must mean that all the tall blonds are Swedish. *A* is right.

14. First cousins

We are dealing with eight children here: five girls and three boys. According to the last two statements, two of the brothers have three children and the last one (Frank's father) has two. From the second and sixth statements, Jack has a daughter Theresa and two sons.

Frank is therefore not Jack's son. Jack's two sons are therefore Ivan and John. According to the fourth statement, Marie has a brother. She is therefore Frank's sister. Isabelle and Ann are therefore sisters to Catherine and all three are Peter's children (fifth statement). Frank and Marie are therefore Paul's children.

15. Nighttime bottle

Let us call i the measure of the inconvenience to the baby from having two bottles instead of one.

By taking the second course of action, the grandmother has one chance in four to make a mistake either one way or the other (too many bottles or not enough). She therefore incurs the following risks:

- i with a probability 1/8 (1/4 × 1/2).
- $2i$ with a probability 1/8.

By taking the first course of action, she has one chance in two of giving the baby too many bottles: hence the risk: i with a probability 1/2.

The likely outcome with the second course of action is therefore $3i/8$ and with the first $i/2$.

The second course of action is thus preferable.

16. Election talk

The brothers always vote in different ways. Hence there are six possibilities:

P_1: Peter for Jones, Jack for Smith, John for Brown

P_2: Peter for Jones, Jack for Brown, John for Smith

P_3: Peter for Brown, Jack for Jones, John for Smith

P_4: Peter for Brown, Jack for Smith, John for Jones

P_5: Peter for Smith, Jack for Brown, John for Jones

P_6: Peter for Smith, Jack for Jones, John for Brown

Six declarations are made during the discussion. It is clear that the first contradicts possibility P_3, the second P_1, the third P_6, the fourth P_4, the fifth P_1, and the sixth P_2. Only the fifth possibility remains. Peter votes for Smith, Jack for Brown, and John for Jones.

17. Journey to Mars

The second cosmonaut asks the Martian: "Are you from this city?" If they are in Mars-Town, it is clear that the Martian will answer "yes," wherever he comes from.

If they are in Mars-City, he will obviously answer "no."

Note: As soon as the cosmonauts find out where they are, their first question will enable them to determine where the Martian who spoke to them comes from. If they are in Mars-Town, the Martian is not lying and indeed lives there. If they are in Mars-City, the Martian is lying and therefore also lives there.

18. Hearing

Alfred says: "Bernard will undoubtedly tell you that he paints houses." This is true. Alfred is therefore not the thief. If he is an accomplice, the statements by Bernard and Charles must be contradictory, which is not the case as both state that Alfred is a piano tuner. Therefore, Alfred, who is neither thief nor accomplice, is innocent. Therefore, he always tells the truth. Thus Charles is an insurance broker, which fits in with what he says himself. Therefore,

Charles can be none other than the accomplice. The accomplice is an insurance broker.

19. Stickup

Since the gangster tried each direction once and once only, there are only six different ways he could have escaped:

1. Right, left, straight ahead
2. Right, straight ahead, left
3. Left, right, straight ahead
4. Left, straight ahead, right
5. Straight ahead, right, left
6. Straight ahead, left, right

But if you examine the first possibility, the three successive directions fit in with the accomplice's statement. In the second possibility, two fit in; in the third, one; in the fourth, two; in the fifth, none fit in; in the sixth, two.

Given that only one statement was true, the only valid possibility is the third. At the first bridge, the gangster turned to the left, at the second to the right, and at the third he went straight ahead.

20. Faulty reporting

Jules and Jim each made four statements; Jack made three. None of the statements made by Jules and Jack match. Jules and Jim made three matching statements, Jack and Jim only one. We can easily deduce that it is Jack who continually lies, Jules who never lies, and Jim who gives one piece of false information—his one statement that is identical to Jack's—that is, when he refers to the steak. His affirmation about the dessert is therefore true. The celebrity ended his meal with strawberry sherbert.

21. Amateur doctor

I advise you to take aspirin and antibiotics and to apply hot compresses to your neck. Aspirin relieves the headache but gives you nausea and upsets your stomach; the antibiotics relieve the nausea but add pains to your knee and give you a stiff neck. The knee is taken care of by the aspirin. As far as the upset stomach and the stiff neck are concerned, the hot compresses will take care of them.

22. Fibs

Let *J* be the proposition "John lies."

Let *P* be the proposition "Peter lies."

John says to Peter: "If *J* is not true, *P* is not true either."

Peter answers: "If *P* is true, then *J* is also true."

Each of these statements indicates that it is not possible to have *J* false and *P* true at the same time. *J* and *P* are therefore synonymous and their authors either are both telling the truth or both fibbing. It is not possible for one to fib and not the other.

23. April Fool!

Let A, B, and C be the three teachers. *A* says to himself: "*B* sees that C is laughing. *B*, however, doesn't know that he has a sign pinned to his back. Therefore, if I didn't have a sign pinned to my back, *B* would be surprised at C laughing and he would deduce that he has a sign pinned to his own back. However, this is not what's happening. *B* keeps on laughing without removing what's pinned to his back. Therefore, it's not possible that I have nothing on my back." Thus, suddenly having understood the situation, *A* stops laughing.

24. Terrorist

We know that of the three statements made, only one is true.

First case: The first statement is true; the Trinians were responsible for the first attempt. The second statement must be false, which is impossible because we know that the Trinians were only responsible for one attempt.

Second case: The second statement is true; the Trinians were not responsible for the January 1 attempt. They were not responsible for the Christmas Day attempt either because the first statement must be false. They must therefore have been responsible for the July 4 attempt. This would make the third statement true, which is impossible.

Third case: The third statement is true. The two first statements are therefore false. The Trinians are responsible for the January 1 attempt. Since the attempt on July 4 was not due to the Bothians, the latter were responsible for the Christmas Day attempt and that on July 4 was the responsibility of the Ruritanians.

The last case is therefore the only one compatible with the given data, and I would report this fact to the police.

25. Conundrum

Lewis Carroll gives us a series of four propositions that we will examine, starting with the last one. "The criminal had a key" (proposition 4). He therefore had an accomplice (proposition 3). We can conclude that he came in a car (proposition 2) and that the witness was not mistaken.

26. Three kings from Orient

There are six possible ways that the three kings could enter the manger:

P_1: Balthazar, Gaspar, and Melchior (or simply *BGM*)

P_2: *BMG*

P_3: *GMB*

P_4: *GBM*

P_5: *MBG*

P_6: *MGB*

Melchior's statements preclude P_4 and P_5.

Balthazar's statements preclude P_2 and P_3.

Gaspar's statements preclude P_6.

P_1 remains. Balthazar came first, followed by Gaspar, then by Melchior.

27. Payday

Brown lies in his first statement and both Smith and Jones lie in their third. Therefore, Brown earns $7000 a month, Smith $9000, and Jones $6,000.

28. High jumps

Let us suppose that none of the three loses her bet. What does this mean? If the first girl was successful in her jump, the second failed (first bet), which means (second bet) that the third was successful but (third bet) that the first failed. This is a contradiction and therefore impossible. There is only one alternative: the first girl missed her jump and the second succeeded (first bet); the third girl missed hers (second bet) but the first girl was successful (third bet). Once again we have a contradiction. The three girls cannot all win their bets.

29. Flowered hats

The answer is obvious and is "no." It is perfectly possible for an Englishwoman to wear a flowered hat without being on a visit to the Empire State Building.

30. Aunt Ernestine

Let i be a measure of the inconvenience of having to go and fetch Aunt Ernestine for nothing. To forget her would mean an inconvenience of value $3i$. Following the first possible course of action, I risk $3i$ once out of two times (because I cannot know whether my wife will remember our aunt). Result: $3i/2$.

Following the second course of action, I risk i once out of two times. Result: $i/2$.

Following the third course of action, I risk $3i$ if Aunt Ernestine does not hear the telephone and is there, that is, if my wife did not go and fetch her. Result: $3i\cdot(1/2)\cdot(2/5) = 3i/5$.

Strategy three will therefore involve a slightly higher risk than strategy two. The latter is therefore the best course of action. I must go and collect Aunt Ernestine, regardless.

31. Apartment complex

Let us examine two possible configurations taken at random:

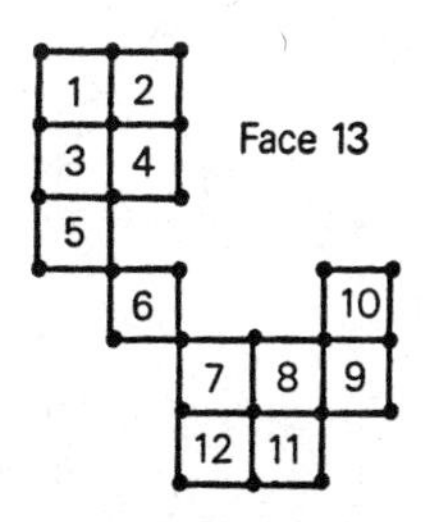

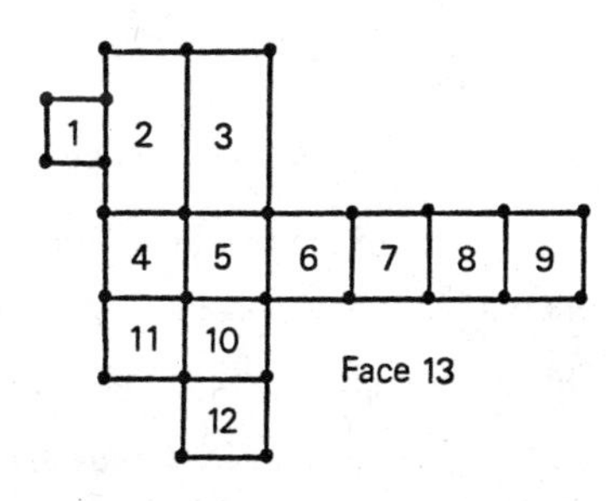

In each configuration, one can see that the number of mailboxes is 26. This, it turns out, is not dependent on the particular layout of the apartment complex. Euler has, in fact, proven the following relationship:

"If in a connected planar figure there are v vertices, s side segments and e enclosed areas, we have $v - s + e = 2$."

Here v is the number of mailboxes, s the number of road segments, and e the number of blocks + 1 (to take into account the outside space). And we do have: $26 - 37 + 13 = 2$.

32. French holiday

Let *E* be the proposition: They visited the Eiffel Tower.

Let *A* be the proposition: They visited the Arch of Triumph.

Let *V* be the proposition: They visited Versailles.

Let *L* be the proposition: They visited the Louvre.

The first nephew said: "*E* and *A* but not *V*."

The second nephew said: "*E* and not *A* and *V* and not *L*."

The third nephew said: "Not *E* but *A*."

If *E* was false, *A* would be true because the first nephew lies only once. This would make the second nephew lie twice, which is impossible. Therefore, *E* is true. Therefore, the third nephew is lying when he says "not *E*." Therefore, *A* is true. Therefore, the first nephew can only be lying when he says "not *V*." Therefore, *V* is true. The second nephew, who lies when he says "not *A*," tells the truth when he says "not *L*." One concludes that Aunt Julia took her nephews up the Eiffel Tower and to visit Versailles and the Arch of Triumph but did not take them to the Louvre.

33. Horse race

Let *X*, *Y*, and *Z* be the following propositions:

- Xavier is one of the first three horses.
- Yellow ribbon is one of the first three horses.
- Zephyr is one of the first three horses.

If the second bet is not lost, it must be that *X* or *Y* is true and that the first bet is lost. However, if the first bet is lost, it must be because *Z* is true and *X* is false simultaneously. We therefore know that if *X* or *Y* is true, *Z* must be true and *X* must be false. Hence *X* must be excluded. Therefore, if *Y* is true, *Z* is true. Zephyr is more likely (or just as likely) as Yellow Ribbon to find itself among the first three horses, whereas Xavier has no chance at all. The best horse to bet on is therefore Zephyr.

34. Prisoners' wives

First solution: The second and the fourth men are telling the truth. The third and the fifth men are then the occasional liars, which is impossible.

Second solution: The first and the fourth men are telling the truth. The others are occasional liars, which is also impossible.

Third solution: The first and the fifth men are telling the truth. The second and the fourth lie all the time and the third lies occasionally.

Conclusion: Ann and Beatrice are blonds. Claire is redheaded. Gertrude and Joanna are brunette.

3. The mysterious relationships among speed, distance, and time

1. New Jersey Turnpike

The New York to Philadelphia turnpike is 150 km long. In my Mercedes, it takes me 25 minutes less time to cover this distance than it takes my wife in her Cadillac. The other day we started at the same time, I from New York, she from Philadelphia. When we met, we observed that the distance (in kilometers) remaining for each of us was equal to the difference in the time (in minutes) left to go if we kept up our usual speeds. How far away in kilometers from Philadelphia were we? How long had we been driving when we met?

2. In Texas

At the same instant that a first airplane leaves Hallelujah City for San Pedro, a second airplane leaves San Pedro for Hallelujah City. They fly at different altitudes at constant speeds, the first being faster than the second. They meet a first time 437 km from San Pedro. Then, one after the other, they land, remain half an hour on the ground to refuel, discharge their passengers, and take on a new load of passengers. Then, one after the other, they take off again toward their original departure points, at the same speeds as the previous leg of their journey. They then meet again 237 km from Hallelujah City. What is the distance separating these two cities?

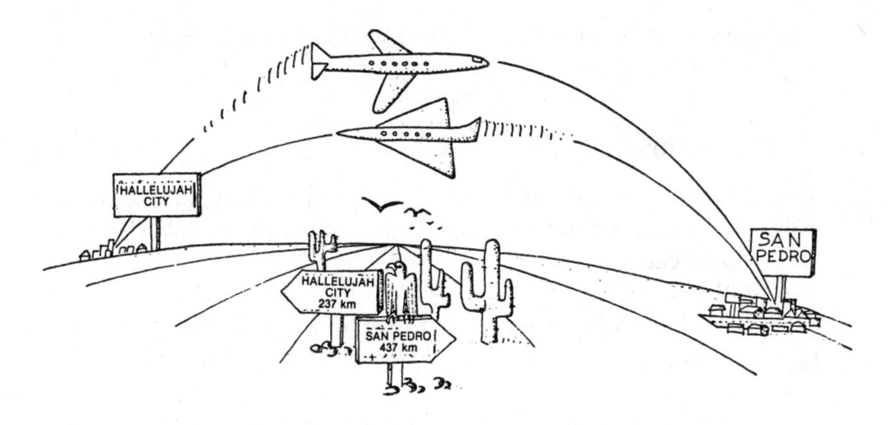

3. Before the Tour de France

A group of cyclists are training for the Tour de France on a small road. Their speed is 35 kilometers per hour (km/h). One of them suddenly leaves the pack at 45 km/h, covers 10 km, does a U turn at the same speed, and returns to the group, all of whom have maintained their previous speed. How much time elapses between the departure and the return of the runaway cyclist?

4. Baltimore-Philadelphia

Two cyclists start at the same time to travel from Baltimore to Philadelphia (distance: approximately 195 km). One of them, whose average speed is 4 km/h higher than the other, arrives 1 hour earlier. What was his speed?

5. Bar flies

At the same instant that Pete was leaving Harry's Bar to go to the Shamrock Tavern, Johnny was leaving the Shamrock Tavern to go to Harry's Bar. They walked at a constant speed. When they met, Pete announced that he had covered 200 meters more than Johnny had. The latter, his mind befuddled by alcohol, took this as a personal insult and started to beat up Pete, who returned the punches. When the fighting stopped, the men embraced in tears, then each continued his original journey, but at half his original speed, as both were slightly hurt. Pete thus arrived at the Shamrock Tavern in 8 minutes while

Johnny took 18 minutes to get to Harry's Bar. How far apart are the two bars?

6. Celebration

To celebrate the Legion of Honor awarded to her husband, a distinguished Parisian woman ordered an assortment of tea cakes from a well-known caterer.

"The party," she confirmed on the telephone, "will start at 6 p.m. sharp. I want you to make your delivery at precisely that time."

"Madame," the caterer replied, "all our delivery trucks leave at fixed times, which conform with the completion of a batch of baking. If the traffic isn't too bad, we usually average 60 km/h and should be at your home at 5:45. But if we run into a traffic jam and our average speed drops to 20 km/h, we won't be able to make your delivery until 6:15 p.m." On the basis of this conversation, can you determine the average speed that would enable the truck to meet the strict schedule requirements of the hostess?

7. The lovelorn cyclist

A young cyclist goes to visit his fiancée to give her a bunch of flowers. He then returns to his starting point. On the journey to see his beloved, encumbered as he is by flowers, he can only average 17 km/h, but on the return journey he averages 23 km/h. What is the cyclist's average speed over the total trip?

8. The sheik and his driver

The sheik Ben Sidi Mohammed drives from his palace to the airport at a constant speed over a splendid expressway built in the middle of the desert. Depending on whether his driver increases his average speed by 20 km/h or reduces it in the same amount, the sheik will gain 2 minutes or lose 3 minutes. What is the distance between the airport and Ben Sidi Mohammed's palace?

9. By car or by cycle?

John and Jules simultaneously leave town *A* to go to town *B*. One travels by car, the other by cycle. After a while it appears that if John had covered three times the distance, he would have only half the remaining distance to go and that if Jules had covered half the distance he would have three times the remaining distance to go. Which man is the cyclist?

10. Subway escalator

I am in the habit of walking up the subway escalator while it is running. I climb 20 steps at my normal pace and it takes me 60 seconds to reach the top, whereas my wife climbs only 16 steps and it takes her 72 seconds to reach the top. If the escalator broke down tomorrow, how many steps would I have to climb?

11. Polar expedition

A polar expedition had a sleigh drawn by five reindeer. After 24 hours, two of them died. This caused the speed of the sleigh to drop to three-fifths of its initial speed, and the explorer arrived at his destination 48 hours later than planned. "What a pity," he cried. "If my reindeer had died 120 km further along the route, I would only have been 24 hours late." How far had the explorer journeyed?

12. A punctual woman

Every Saturday afternoon, I play tennis with my friend Philip from 3 to 4 o'clock. My wife picks me up in her car at 4:10 precisely. The other day, however, Philip had the flu. I had not been alerted, so I went to the court as usual. At 3:05, realizing that Philip was not going to show up, I left the court and started walking home. After a while, I met my wife, who was on her way to fetch me. I got into the car and we returned home, arriving 10 minutes earlier than usual. Can you determine the ratio of my speed on foot to my wife's speed by car?

13. The big drum and the cub scout

A parade, 250 meters long, was organized for National Boy Scouts Week. The cub scouts were up in front, the band at the tail end. Soon after moving off, one of the cub scouts, who was carrying the flag, remembered that he had left his cap with the girl who played the big drum in the last row of the band. He ran to the tail of the column at 10 km/h to retrieve his cap and returned to his place in the ranks 3 minutes 18 seconds later. At what speed was the parade progressing?

14. Janet and Suzy

Janet and Suzy dive simultaneously, from the same spot, into a narrow river. Janet swims upstream, Suzy downstream. However, Suzy forgets to remove a necklace of wooden beads, which comes off as she hits the water and is carried off by the current. A quarter of an hour later, the swimmers turn around. Knowing that both swim at the same speed when there is no current, can you determine whether Suzy will be able to find her necklace before she meets up with Janet?

15. Mexican subway

In a tunnel of Mexico City's subway, a workman walks along the track between two stations that are some distance apart. Every 6 minutes a train overtakes him. Every 5 minutes a train comes by from the opposite direction. The workman walks at a constant speed and the trains appear at regular intervals (moving at constant speed in either direction). After a while, the workman stops to repair a switch. A train passes by. How long will it be before another train appears coming from the same direction?

16. Navigation

A boat runs on a river with a current of 3 km/h. It travels in one direction, immediately turns around, and returns to its starting point 6 hours later, having covered 36 km on the chart. At what speed was it traveling through the water?

17. Washington/Ocean City

Mr. and Mrs. Smith travel to Ocean City, where they own a summer home. Each drives his or her own car. They start off together and arrive together. Mr. Smith, however, stops on the way for a time equal to one-third of the time that Mrs. Smith drives, whereas Mrs. Smith stops for a time equal to one-fourth of the time Mr. Smith drives. What is the ratio of the average driving speeds of husband and wife?

18. Errands

At the same time that the butcher asks his son to go and buy bread at the bakery, the baker asks his son to go and buy meat at the butcher shop. The two boys walk toward each other at a constant speed. When they cross, the butcher's son has covered 500 meters more than the baker's son. The former has another 10 minutes to go, whereas the latter has another 22.5 minutes to go. What is the distance between the bakery and the butcher shop?

19. Picnic

At 9 a.m. Paul leaves to ride his bicycle from A to B at a speed of 15 km/h. At 10 a.m., Peter does the same, going from B to A at 20 km/h. They meet halfway to have a picnic. What time is it then?

20. The diver

A swimmer dives from a height of 10 meters. He passes the 5-meter board at a speed v and hits the water at a speed $2v$. Any comment?

21. Rue Sainte-Catherine

I was shopping with my mother-in-law on a trip to France. We were walking slowly (3 km/h) along the Rue Sainte-Catherine in Bordeaux, which is absolutely straight. Suddenly, I remembered that I had an urgent letter to mail. Noticing a mailbox a bit farther along the street, I left my mother-in-law strolling along and dashed off, still heading in the same direction, at 5 km/h. I returned at the same speed and noticed that I had only been separated from my mother-in-law for 3 minutes. How far were we from the mailbox when I dashed off?

22. Early closing

Frances gets out of school every day at 3 and her mother picks her up by car. One snowy day, the pupils are dismissed earlier than usual. Frances starts walking home and, after a quarter of an hour, meets her mother coming to pick her up at the normal time. Frances gets in the car, her mother turns the car around, and they arrive back at the house 10 minutes earlier than usual. At what time was Frances dismissed from school?

23. Cross-country skiing

A cross-country skier hesitates between two trails. Both trails are of equal length but the first is completely level, whereas the second is half uphill and half downhill. Our skier knows that he is three times slower going uphill than on the level but also that he is three times faster going downhill than on the flat. He quickly decides it will take him the same amount of time to do either trail. Is he right?

24. The two whales

In the middle of the Atlantic Ocean, two whales were swimming peacefully on a straight-line course at 6 km/h. One of the whales decided to speed up a little and moved off at 10 km/h, still heading in the same direction. It then decided to reverse its course and return to its friend, which had changed neither speed nor direction. Knowing that the two whales parted at 9:15 a.m. and were reunited at 10 a.m., what time was it when the faster whale turned back?

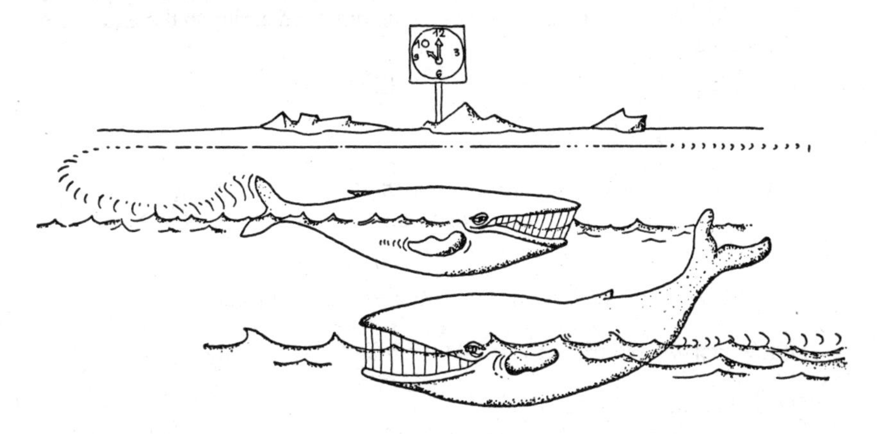

25. Subsonic and supersonic

Pilots Smith and Jones take off simultaneously from Heathrow to fly to Kennedy. One of them flies a supersonic plane, the other a subsonic aircraft. After a certain time has elapsed, it appears that if the plane piloted by Smith had covered twice the distance that it in fact has, the remaining journey would take one and a half times less time. Also, if the plane piloted by Jones had covered one and a half times less distance, the remainder of the journey would take twice as long. Which pilot is piloting the supersonic aircraft?

26. Moving sidewalk

When I take the sidewalk in the right direction, it takes me 6 seconds to travel its length, walking at my normal speed. Going in the opposite direction, it takes me 6 minutes, which is the time I normally take to cover 500 meters. What is the length of the moving sidewalk?

One step at a time

1. New Jersey Turnpike

1. Let x be the distance from Philadelphia to the point where my wife and I met (in kilometers).

Distance to New York: $150 - x$.

Difference between these two distances: $150 - 2x$, which must be equal to 25 km since 25 is also the difference in the times remaining to the end of the trip. Hence

$$150 - 2x = 25$$

$$x = 62.5 \text{ km}$$

2. The Mercedes has therefore covered $150 - 62.5 = 87.5$ km at a speed of V (in km/h). The Cadillac has covered 62.5 km at a speed of v. Hence the equations

$$62.5/v = 87.5/V \qquad \text{and} \qquad 150/v - 150/V = 25/60$$

from which we can extract $V = 144$ km/h and $v = 103$ km/h.

Hence the required time is

$$8.75/V = 36 \text{ minutes and } 28 \text{ seconds.}$$

2. In Texas

Let d be the distance between Hallelujah City and San Pedro. When they first cross, the aircraft have covered a total distance d. When they cross for the second time, they have covered a total distance $3d$. Since both aircraft spend the same length of time on the ground, each aircraft has covered, at the time of the second crossing, three times the distance covered to the first crossing. Hence for the first aircraft we have

$$d - 437 = 1/3(2d - 237)$$

Hence

$$d = 974$$

Hallelujah City and San Pedro are 974 km apart.

3. *Before the Tour de France*

Let x be the time that elapses between the departure and the return of the runaway cyclist. The distance he covers is $45x$ km, during which time the pack covers $35x$ km. The sum of these two distances, is of course, equal to the distance covered by the solitary cyclist on his outbound leg, added to his return leg, added to the distance covered by the pack during this time. The total distance is also twice the distance covered by the solitary cyclist on his outbound leg. Hence

$$45x + 35x = 20$$

Hence $x = 1/4$.

Fifteen minutes elapses between the departure and the return of the cyclist.

4. *Baltimore-Philadelphia*

Let v be the required speed in km/h. Let t be the time (in hours) taken by the faster cyclist.

Speed of the slower cyclist: $v - 4$.

Corresponding time: $t + 1$.

We therefore have the following relationship:

$$195 = vt = (v - 4)(t + 1)$$

Hence $t = 195/v$ and $v - 4t - 4 = 0$.

Hence $v^2 - 4v - 4(195) = 0$, which is a quadratic with one positive root; hence, $v = 30$ km/h, which is the required speed.

Time taken by the faster cyclist: $195/30 = 6\frac{1}{2}$ hours.

Time taken by the slower cyclist: $195/26 = 7\frac{1}{2}$ hours.

5. *Bar flies*

Let the units used be meters and minutes.

Let x be the distance between the two bars.

Let d be the distance covered by Pete when he met Johnny.

Let V be Pete's speed and v be Johnny's speed before the fight.

The sum of the distances covered at the time they met is x:

$$d + (d - 200) = x$$
$$\text{Hence } x = 2d - 200 \qquad (1)$$

When they met, each had walked for the same length of time:

$$d/V = (d - 200)v \qquad (2)$$

After the fight, Pete walked for 8 minutes:

$$(d - 200)(V/2) = 8$$

Hence $V = (d - 200)/4$ (3)

and Johnny for 18 minutes:

$$d/(v/2) = 18$$

Hence $v = d/9$ (4)

By substituting the values for V and v into equation (2), we have

$$4d/(d - 200) = 9(d - 200)/d$$

Hence

$$5d^2 - 3600d + 360{,}000 = 0$$

Solving this quadratic in the classic way yields

$$d = 600 \text{ or } 120 \quad \text{and} \quad x = 2d - 200 = 1000 \text{ or } 40$$

However, d must be smaller than x, so only one solution is possible:

$$d = 600 \quad \text{and} \quad x = 1000$$

The distance between the two bars is exactly 1000 meters.

6. Celebration

Let v be the unknown speed, t the delivery time, and d the distance (in km) to be covered. This gives us the following equations:

$$t = d/v$$
$$t + 1/4 = d/20 \rightarrow 20t + 5 = d$$
$$t - 1/4 = d/60 \rightarrow 60t - 15 = d$$

Hence

$$40t - 20 = 0; \text{ thus}$$
$$t = 1/2,\ d = 15, \text{ and } v = 30.$$

The delivery van must run at an average speed of 30 km/h.

7. The lovelorn cyclist

Let d be the distance to be covered to the home of the cyclist's fiancée.

Time taken for the outgoing trip: $d/17$.

Time taken for the return trip: $d/23$.

Total time: $0.102d$.

The average speed is therefore

$$2d/(0.102d) = 19.6 \text{ km/h}$$

8. The sheik and his driver

Let l be the unknown distance (in km).

Let v be the speed of the sheik's car (in km/h).

Let t be the time taken to go to the airport (in hours).

We have the following three equations:

$$v = l/t$$
$$v + 20 = l/(t - 2/60)$$
$$v - 20 = l/(t + 3/60)$$

Eliminating l, we have

$$(v + 20)(t - 2/60) = vt \rightarrow -v/30 + 20t - 2/3 = 0$$
$$(v - 20)(t + 3/60) = vt \rightarrow v/20 - 20t - 1 = 0$$

Hence $v = 100$ and $t = 1/5 = 12$ minutes; thus, $l = vt = 20$.

The distance between the airport and Ben Sidi Mohammed's palace is 20 km.

9. *By car or by cycle?*

Let x be the distance covered by John, and x' the distance that remains to be covered.

Let y be the distance covered by Jules, and y' the distance that remains to be covered.

We have

$$x + x' = 3x + x'/2 \qquad \text{and} \qquad y + y' = y/2 + 3y'$$

Hence

$$x = x'/4 \qquad \text{and } y = 4y'$$

John has therefore covered a fourth of the distance that remains, whereas Jules has covered four times the distance that remains. The latter is therefore in a car and the cyclist is John.

10. *Subway escalator*

Let x be the unknown number of steps. When I ride the escalator, I cover $x - 20$ steps in 60 seconds. When my wife uses the escalator, she covers $x - 16$ steps in 72 seconds. The escalator therefore moves at a rate of 4 steps every 12 seconds or 20 steps in 60 seconds. Its total height is therefore the sum of these 20 steps and of the additional 20 steps I climb: a total of 40 steps.

11. *Polar expedition*

An additional distance of 120 km covered by the two reindeer corresponds to a 24-hour reduction in delayed time of arrival. Therefore, in order to make up the 48 hours of delay, it would have been necessary for the reindeer to survive another 240 km.

Let v be the speed at the beginning of the trip. The time taken for the explorer to reach his destination after the two reindeer died can be written as:

$$48 + 240/v \qquad \text{or} \qquad 5/3 \times 240/v$$

Hence $48 + 240/v = 400/v$; thus $v = 10/3$ km/h.

Before the reindeer died, the explorer had covered (in 24 hours) $24 \times 10/3 = 80$ km. The total distance to be covered was therefore $80 + 240 = 320$ km.

12. *A punctual woman*

My wife drove for 10 minutes less than she usually did, 5 minutes less in each direction. I therefore met her at 4:05 p.m. and not at 4:10 p.m. as usual. I had walked for 60 minutes, covering the distance that my wife would have covered in 5 minutes. My speed on foot was therefore one-twelfth of her speed by car.

13. The big drum and the cub scout

Let x be the speed of the parade. In going to fetch his cap, it was as though the cub scout covered the 250 meters at a speed of $10 + x$. Returning to his place in the parade, it was as though he covered 250 meters at a speed of $10 - x$ (kilometers/hours).

The total elapsed time was therefore

$$\frac{250}{1000} \cdot \frac{1}{10+x} + \frac{250}{1000} \cdot \frac{1}{10-x} = \frac{3}{60} + \frac{18}{3600}$$

(the units are kilometers, hours, and kilometers/hour). Expanding this equation, we have

$$(10 - x) + (10 + x) = 4(10 + x)(10 - x)(3/60 + 18/3600)$$

or

$$20 = 4(100 - x^2)(198/3600)$$

Hence $x^2 = 100/11$ and $x \simeq 3$.

The parade was advancing at approximately 3 km/h.

14. Janet and Suzy

The two swimmers move at their own speeds relative to the water of the river. The floating wooden necklace is stationary relative to the water of the river. The swimmers cover equal distances and arrive simultaneously half an hour later at the floating necklace. Suzy therefore recovers her necklace at the same time as she meets up with Janet.

15. Mexican subway

Let x be the time between the passage of two trains.

Let l be the distance between two successive trains.

Let V be the speed of the trains.

Let v be the speed of the workman.

Any train overtaking the workman has a relative velocity of $V - v$.

Any train that comes in the other direction has a relative velocity of $V + v$.

Hence we have the set of equations

$$5 = l/(V + v)$$

$$6 = l/(V - v)$$

$$x = l/V$$

Hence $x = 60/11 = 5$ minutes 27 seconds.

16. Navigation

Let x be the speed of the boat in km/h.

When the boat is running with the current, the effective speed will be $(x + 3)$.

When the boat is moving against the current, the effective speed will be $(x - 3)$.

Since the total travel time is 6 hours, we have

$$18/(x + 3) + 18/(x - 3) = 6$$

Hence $x^2 - 6x - 9 = 0$.

Solving this quadratic for x yields

$$x = 3 + 3\sqrt{2} = 7.24 \text{ km/h}$$

17. Washington/Ocean City

Let a and a' be the duration of the stops made by Mr. and Mrs. Smith, respectively.

Let r and r' be the total driving times.

Since Mr. and Mrs. Smith start together and finish together, we have

$$a + r = a' + r'$$

But we also know that

$$a' = r/3 \text{ and } a = r'/4$$

Hence $r'/4 + r = r/3 + r'$; thus $r = 9r'/8$.

Mrs. Smith therefore drives at eight-ninths of the average speed of Mr. Smith.

18. Errands

Let us use as units kilometers, hours, and kilometers/hour.

Let x be the distance that the baker's son has covered when the two boys pass each other, v_2 his speed, and v_1 the speed of the butcher's son.

Both boys walk for the same time before meeting; hence

$$(x + 0.5)/v_1 = x/v_2$$

The information given concerning the times it takes the two boys to reach their destinations can be written

$$x = 10/60\ v_1$$

$$x + 0.5 = 22.5/60\ v_2$$

Hence $v_1 = 6x$ and $v_2 = 8/3\ (x + 0.5)$.

Substituting in the first equation yields

$$(x + 0.5)/6x = x/(8\cdot 3)(x + 0.5)$$

Hence $10x^2 - 8x - 2 = 0$, of which the only positive root is $x = 1$.

The baker's son walks for 1 km before meeting the butcher's son, who therefore has covered 1.5 km.

The distance between the butcher's shop and the bakery in this small village is therefore 2.5 km.

19. Picnic

Let x be half the distance between A and B.

Time taken by Paul to reach the meeting place: $x/15$.

Time taken by Peter: $x/20$.

Since Peter left 45 minutes after Paul, we can write

$$x/15 = x/20 + 0.75$$

Hence $x = 45$.

Paul took 3 hours to reach the meeting place; they therefore had their picnic at noon.

20. *The diver*

The feat described is impossible. The change in kinetic energy between the 5- and 10-meter marks should be the same as the change in kinetic energy between 0 and 5 meters:

$$(v_5)^2 - 0^2 = (v_{10})^2 - (v_5)^2$$

$$\text{Hence } (v_{10})^2 = 2(v_5)^2.$$

In other words, the speed of arrival at water level should be 1.414 times the speed at half the height of the 10-meter board.

21. *Rue Sainte-Catherine*

Let us use as units meters and minutes.

Own speed: $5000/60 = 250/3$.

Mother-in-law's speed: $3000/60 = 50$.

Let S be the place where we split, R the place where we met again, and B the mailbox.

Distance covered by my mother-in-law in 3 minutes:

$$SR = 50 \times 3 = 150$$

Distance covered by me:

$$SB + BR = SR + 2BR = (250/3) \times 3 = 250$$

Hence $2BR = 100$ and $BR = 50$; thus $SB = SR + RB = 200$.

When I dashed off, we were 200 m from the mailbox.

22. *Early closing*

The mother's trip has been cut short by 10 minutes: 5 minutes that would have been taken to reach the school and 5 minutes to return to the spot where she picked up her daughter. She therefore met Frances 5 minutes earlier than usual. Frances had been walking for 15 minutes. She had therefore been let out from school 20 minutes early, that is, at twenty to three.

23. Cross-country skiing

Let t be the time taken by the skier over the level trail. Over the uphill part of the second trail (which is half the length of the whole trail), the skier's speed is reduced to one-third and the time taken is therefore $3t/2$. This is already greater than t. Since the time taken on the downhill leg has still to be taken into account, it is obvious that the skier will take longer to cover the second trail than the first one. Therefore, his conclusion is wrong.

24. The two whales

Let t_1 be the time that elapsed between the moment when the two whales separated and the moment when the faster whale turned around.

Let t_2 be the time that elapsed between the moment when the faster whale turned around and the moment when it rejoined its slower friend.

$t_1 + t_2 = 3/4$ if we take the hour as the unit of time.

Distance between the point where the whales separated and the point where they met again: $6(t_1 + t_2) = 10t_1 - 10t_2$.

We therefore have two equations with two unknowns:

$$4t_1 + 4t_2 = 3 \qquad \text{and} \qquad 4t_1 - 16t_2 = 0$$

Hence $t_2 = 3/20 = 9$ minutes.

The faster whale turned around at 9:51 a.m.

25. Subsonic and supersonic

Let x be the distance covered by Smith and x' the distance remaining.

Let y be the distance covered by Jones and y' the distance remaining.

The total length of the journey can be written in four different ways:

$$x = x' \qquad 2x + (x'/1.5) \qquad y + y' \qquad (y/1.5) + 2y'$$

From the first two expressions: $x = x'/3$.

From the last two expressions: $y = 3y'$.

Smith has therefore covered one-fourth of the total journey and Jones has covered three-fourths. Jones is therefore piloting the supersonic aircraft.

26. Moving sidewalk

Let v be the speed of the moving sidewalk, in meters per minute.

Let V be my own speed, also in meters per minute.

Let x be the length of the moving sidewalk, in meters.

We have

$$x/(V + v) = 1/10 \qquad \text{and} \qquad x/(V - v) = 6$$

Furthermore, we know that $V = 500/6$.

Hence

$$x/(500/6 + v) = 1/10 \qquad \text{and} \qquad x/(500/6 - v) = 6$$

Hence, by eliminating v: $x = 1000/61 = 16.39 =$ the length of the moving sidewalk.

4. Geometry, the good old way

1. Bisector

In any triangle, the bisector of one of the angles is also the bisector of the angle formed by the height (or altitude, perpendicular from a vertex to the opposide side) and the diameter of the circumscribed circle passing through the vertex in question. Why is this so?

2. North Sea disaster

In order to prospect for oil, a derrick, set on a heavy concrete base, is placed in the North Sea. When the sea is calm, the portion of the derrick that is above water stands 40 meters high. During a violent storm, the derrick tips over on its base. The disaster is filmed from a nearby rig. The photographs show the tip of the derrick disappearing underwater 84 meters from the point where the derrick originally emerged from the sea. How deep is the water at the original site? (*Note:* Wave height is not a consideration.)

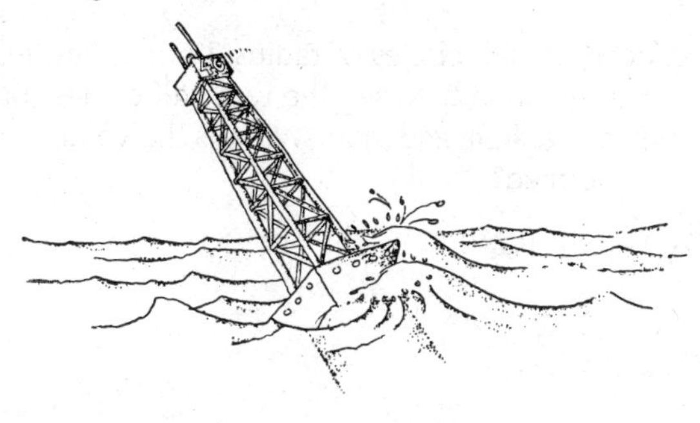

3. A quiet spot

I recently bought a charming piece of land in the country at an excellent price. But the area turned out to be noisier than I'd expected because my property lies within three segments of straight roads of equal length: a parkway, a turnpike, and a trucking route. I plan to build my house at a location such that the sum of the distances to the three roadways is maximized. Where will that point be?

4. Construction

Do you know how to draw two parallel lines, a given distance l apart, passing through the two given points A and B ($l <$ distance AB)?

5. Pencil, eraser, T-square, and compass

Gather the following items: a sheet of white paper, an eraser, a T-square, a compass, and a pencil. Can you now precisely determine two segments the sum of whose lengths is equal to that of the pencil and the square root of whose product is that of the eraser?

6. Lost in the desert

Two men in a Jeep are lost in the midst of a desert. For the last few hours they have observed with the help of elaborate equipment that the sum of the squares of their distances from the oasis Sidi-Ben and the oasis Ben-Sidi equals twice their distance from oasis Sidi-Sidi squared. "We are clearly on a perpendicular to the track that joins Sidi-Sidi to the middle of the track that joins Sidi-Ben to Ben-Sidi," announces the driver. Is he correct?

7. Cutouts

Draw two concentric half-circles of radius 2 and 4 cm, respectively. Cut out the area included between the two half-circles, butt the two straight segments together, and glue. What is the volume of the truncated cone thus formed?

8. Dublin, 1956

The Irishman Burlet proved in 1856 that any right-angle triangle has the same area as the rectangle with sides equal to the two segments formed on the hypotenuse by the point of contact of the inscribed circle. How did he do it?

9. Water, oil, and mercury

Mercury (density 13.59), water (density 1), and oil (density 0.915) are poured in that order into a cone-shaped glass. The three liquids fill the glass in three equal layers without intermixing. Which layer weighs the most?

10. Sheik

A wealthy sheik lolls on the desert sands contemplating his latest jewel, a small sheet of gold incrusted with diamonds, shaped like an equilateral triangle. As he lovingly turns his treasure from side to side, he suddenly notices that its shadow has the form of a right-angle triangle whose hypotenuse equals the actual length of each side of the jewel. What is the angle formed by the jewel's plane and the horizontal desert floor, he wonders. Before finding the answer, he falls asleep under the noonday sun. Can you solve his problem for him by calculating the cosine of the angle in question?

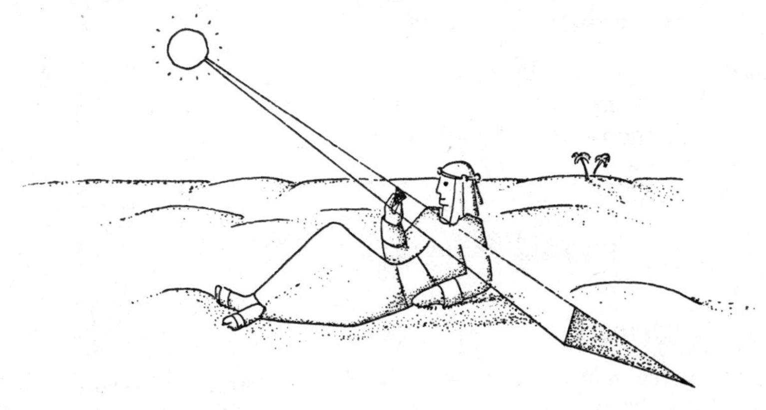

11. Windshield wipers

A windshield is swept by two wipers, each with a given length L. Each wiper sweeps a half-circle. The distance between the centers of the half-circles is L. What is the total area swept?

12. Figure

Five points are laid out on a plane in such a way that no three are aligned. The points are then connected by twos, resulting in straight lines. Supposing that no two of these straight lines are parallel, how many points of intersection will there be in addition to the original five?

13. Christmas pie

On Christmas day a mother serves her children a pie containing a good-luck penny. She first starts to cut the pie in half along a diameter but stops when her knife hits the penny. She then makes a second straight cut, which forms a 45° angle with the first cut. In so doing, she touches the penny again but continues and completes the cut. Her children, who think more about math than about food, notice that the sum of the squares of the segments into which the second cut is divided by the penny equals twice the square of the pie's radius. Do you think this is coincidental?

14. Revolutionary geometry

Let a triangle ABC be right angled at B. Let M be the point of the hypotenuse equidistant from the two sides of the triangle. Can you then find the value of the following?

$$E = \sqrt{1830}\,(AC - \sqrt{AB^2 + BC^2}) + 1789 - \frac{1/AB + 1/BC - \sqrt{2}/BM}{(1848)^3}$$

15. High mass

The main window of a modern church is shaped like a circle, 2 meters in diameter. It is divided by two perpendicular straight lines which intersect at a point 50 cm from the center of the window. During high mass, a daydreaming parishioner calculates the sum of the squares of the lengths of the two segments of the cross. "How amusing," he says to himself. "I've come up with as many meters squared as there are capital sins." How can this be?

16. Legacy

A farmer owns a huge field shaped like a parallelogram, which we will call *ABCD*. Someplace within the field at point *O* is a well. Believing that he is about to die, the farmer bequeaths two triangular fields *AOB* and *OCB* to his son Peter and the rest of the land to his son John. They receive joint ownership of the well. Knowing that the length of *AB* is greather than that of *BC*, which of the two brothers do you think got the larger share?

17. Hexagon

Consider a hexagon and tell me how many diagonals there are. If you find this too easy, extend the six sides and state the total number of points of intersection of all these straight lines (sides and diagonals).

18. Intersections

Consider any quadrilateral. Show that the two segments that join the center points of opposite sides, as well as those that join the center points of the diagonals, meet in their own center points.

19. Italy, 1678

Perhaps in your youth you were taught the theorem of John de Ceva, a mathematician who lived in the seventeenth century. This is what he said: Let *P* be a point picked at random inside a triangle *ABC*. The extended lines *AP, BP,* and *CP* cut the triangle at *X, Y,* and *Z*. We then have

$$\frac{BX}{CX} \cdot \frac{CY}{AY} \cdot \frac{AZ}{BZ} = 1$$

Can you prove this?

20. Medians

Is it possible that the sum of the lengths of the medians of a triangle is less than three-fourths of its perimeter?

21. Parallelogram

A parallelogram of fixed shape *ABCD* moves in its plane in such a way that two adjacent sides *AB* and *AD* (extended) pass by two points *M* and *N*. Can you prove that the diagonal *AC* also passes by a fixed point?

22. Lighthouse

How far is the horizon from the top of a 125.7-meter-high lighthouse? (The earth can be considered spherical with a circumference of 40,000 km.)

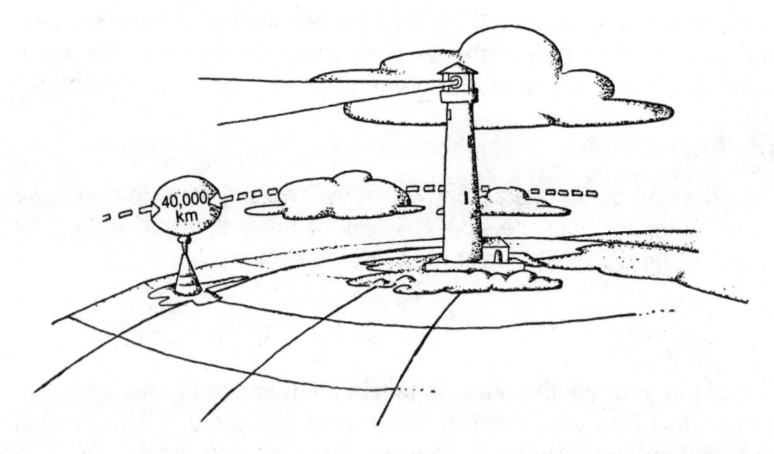

23. Polygons

Divide a circle into n equal arcs. Join every pth point until you return to the initial point. How many sides are there to the polygon thus formed? Check the general formula for $n = 10$ with, successively, $p = 2$, $p = 3$, and $p = 4$.

24. Walking the dog

Mr. and Mrs. Jones decide to take their dog for a walk. Since they both want to hold the dog's leash, they end up by using two separate leashes on the poor animal. Each leash is 1 meter long. Knowing that Mr. and Mrs. Jones always walk 1 meter apart, what is the area in which the dog is free to wander?

25. Quadrilateral

A quadrilateral inscribed in a circle presents the following peculiarity: one of its diagonals is a diameter. What can you say of the projections, on the other diagonal, of the four sides of this quadrilateral?

26. The ramparts

A small town is surrounded by ramparts in the form of a semicircle. There are three gates. The first two of these, P_1 and P_2, are at the ends of the straight side. The third gate, P_3, is at a random point on the semicircle. From each gate, roads, tangent to the semicircle, lead away from the town. These roads meet at two crossroads. Another road follows the straight part of the walls.

A cyclist pedals around the town at a constant speed, following the roads described above. He notes that if he squares the time he takes to go from P_1 to P_2, he obtains the product of the time taken to cycle from P_1 to P_3 by the time taken to cycle from P_3 to P_2. How do you explain this?

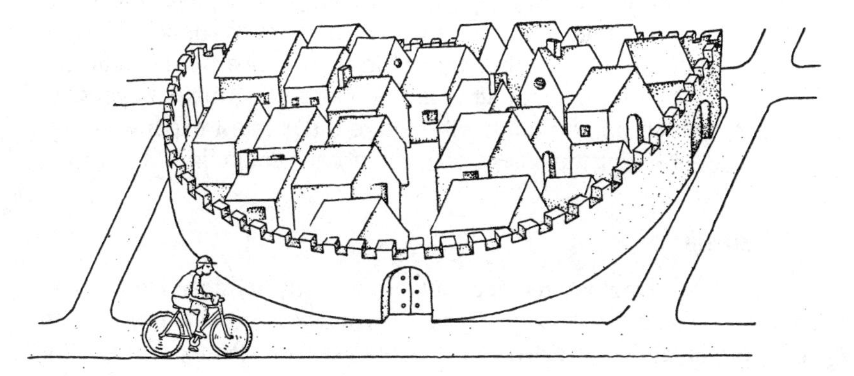

27. Equal areas

Consider an isosceles triangle *ABC* with a right angle at *A*. Draw the circumscribed half-circle, then the arc of the circle tangent to *AB* at *B* and to *AC* at *C*, inside the triangle. The area comprised between the arc and the semicircle is equal to that of the triangle itself. Why is this?

28. Tangency

Two circles are externally tangent. Two straight lines are drawn through their point of contact. These lines cut the circles in four other points. What type of quadrilateral is generated when these four points are joined?

29. Triangle

Can you construct a triangle knowing its perimeter, one angle, and the altitude from the corresponding vertex?

30. Right-angle triangle

In any right-angle triangle, the sum of the sides of the right angle is equal to the sum of the diameters of the inscribed and circumscribed circles. Why?

31. If the earth were an orange

Wrap a red string around an orange. Add enough length to the string to encircle the orange, staying 1 meter away from its surface. Wrap a blue string around the earth (assumed to be spherical). Add sufficient length to the string so that it goes around the earth 1 meter beyond its surface. Which do you think will require the greatest expansion: the red string encircling the orange or the blue string encircling the earth?

32. Siesta

Albert dozes, lazily stretched out on a sunny terrace. He amuses himself by studying his right hand: "My index finger makes an angle of approximately 20° with my middle finger," he says to himself. "If I maintain this spacing, can I rotate my hand in such a way that the shadows cast by my fingers onto the ground are at right angles?" A minute later, having found the answer, he falls fast asleep. Would you have gone to sleep that quickly had you been in his place?

33. Statues

Three statues of historical figures are located in a French park. They are placed along a straight path which passes successively over four small, unequally spaced bridges. One statue stands between each set of bridges in the following order: Vercingetorix, Charlemagne, and Henri IV. A statue of Louis XIV is later erected in the middle of the interval between the first and fourth bridges. The city fathers subsequently plan to add a fifth statue, personifying Napoleon, to the park. Where would you recommend placing it?

- Halfway between Vercingetorix and Henri IV?
- Halfway between Charlemagne and Louis XIV?

34. Street talk

The main street of a French town cuts the Rue General de Gaulle at a right angle in front of the town hall. The Catholic church is located on the main street and the Protestant church faces on the Rue General de Gaulle. The school is located on a third street, one that joins the two churches, equidistant (500 meters), as the crow flies, from each of the other two streets. All the streets are straight. A schoolteacher asks his pupils one day to compute the sum of the reciprocals of the distances in meters from the town hall to each of the two churches. They know neither trigonometry nor analytical geometry. What is the answer that they must give their teacher?

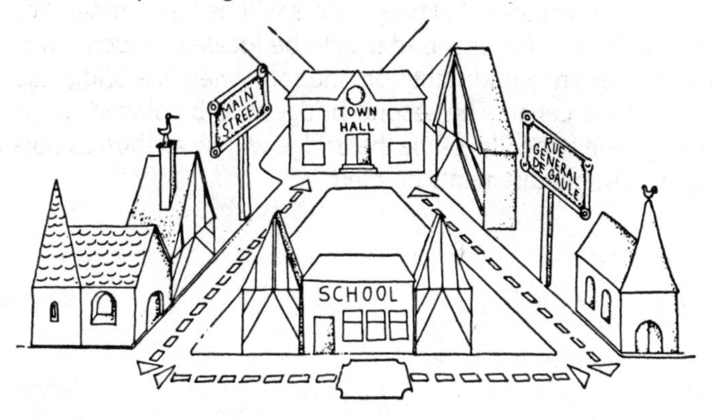

35. View from a plane

From a plane's window, I can see part of an island, part of a cloud, and a bit of sea. Knowing that the cloud occupies half the view and that the cloud hides one-fourth of the island, which appears to fill one-fourth of the view, what is the proportion of sea hidden by the cloud as seen from this window?

36. Pedestrian zone

A pedestrian zone includes seven small straight streets. Is it possible that each street crosses no more and no less than three of the others?

37. Two circles

Two circles are tangent individually at a point *A*. Let *B* be the point diametrically opposite *A* on the larger circle. Chord *BD* is tangent to the smaller circle at *C*. It will be found that *AC* is the bisector of $\widehat{BAD}$. Why?

38. Crescents

Consider a triangle *ABC*, right angled at *A*. Draw the circumscribed half-circle, then the two half-circles constructed on *AB* and *AC* as diameters and outside the triangle *ABC*. The sum of the areas of the two crescents thus formed is equal to that of the triangle *ABC*. Why is that?

39. The two bridges

A large, straight canal passes between two towns, closer to one than to the other. It is decided that two bridges will be built perpendicularly to the canal banks. The first bridge is to be located in such a way that the two villages are equidistant from the spot where the bridge touches their side of the canal. The second bridge is to be placed so that the road joining the two villages via the bridge will be as short as possible. How can these requirements be met?

40. Three perpendiculars

Can you demonstrate that the sum of the lengths of the three perpendiculars drawn onto the sides from a point inside an equilateral triangle is independent of the position of that point?

41. The four suits of armor

A museum displays a suit of armor in each corner of a small square room measuring 3.414 meters per side. The museum director decides that the armor could be better exhibited if each of the four suits were placed in a triangular niche (one per corner) designed in such a way that the remainder of the room would become an octagon with equal sides. The carpenter hired to do the work must determine what the dimensions of the niches should be. Can you help him?

Some have it, some don't: the knack for geometry

1. Bisector

Let ABC be any triangle, O the center of the circumscribed circle, A' the point diametrically opposite A, and H the foot of the perpendicular drawn from A onto BC.

This gives us

$$\widehat{HAC} = \pi/2 - \widehat{BCA} \quad \text{and} \quad \widehat{BAA'} = \pi/2 - BA'A$$

But $\widehat{BCA} = \widehat{BA'A}$ because they intercept the same arc.

Hence $\widehat{BAA'} = \widehat{HAC}$ because they are the complement of two equal angles.

Therefore, angle $\widehat{A'AH}$ has the same bisector as $\widehat{BAC}$.

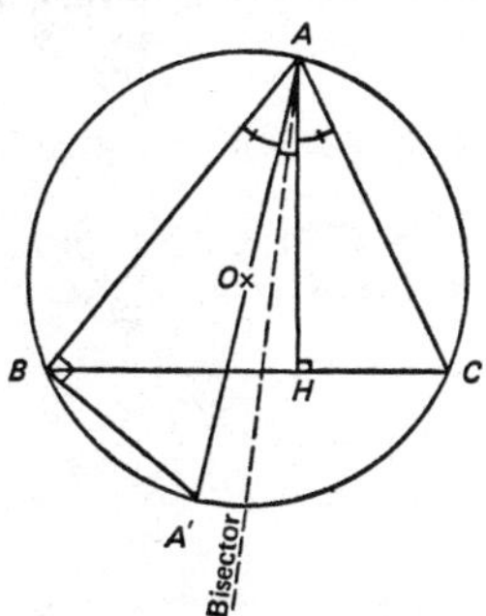

2. *North Sea disaster*

Let x be the unknown depth (in meters).

Let O be the foot of the derrick.

Let H be the point of emergence.

Let A be the point where the top of the derrick was before the disaster.

Let B be the point where the top of the derrick fell into the sea.

Draw a circle of center O and radius OA. Extend BH to B' and AO to A'. We therefore have

$$HA \cdot HA' = HB \cdot HB'$$

In other words, $40(40 + 2x) = 84^2$.

From this we determine that $2x = 84^2/40 - 40$ or $x = 68.2$.

The depth of the sea at the original site is 68.2 meters.

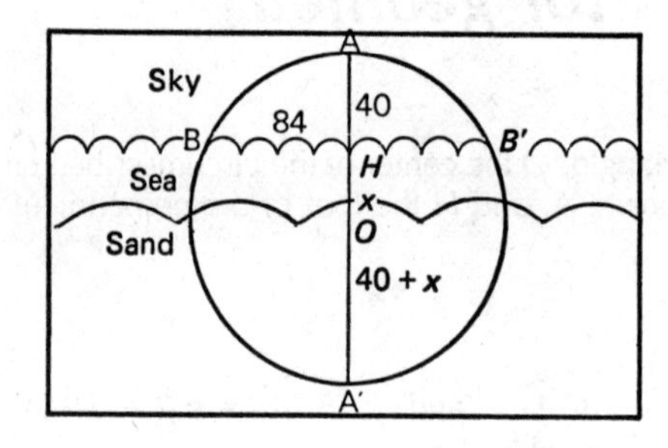

3. *A quiet spot*

My piece of land is an equilateral triangle. I can locate my house at any point within this triangle since from any such point the sum of the distances to the three sides will be the same. This is so because if ABC is the triangle and M a random point inside it,

$$\text{area } ABC = \text{area } AMB + \text{area } BMC + \text{area } CMA$$

This can be written: area $ABC = (1/2) \times AB \times$ distance from M to AB + $(1/2) \times BC \times$ distance from M to BC + $(1/2) \times CA \times$ distance from M to CA. Therefore,

$$\text{sum of the three distances} = \frac{2 \times \text{area } ABC}{\text{length of the side of the triangle}}$$

and this is obviously independent of the position of M.

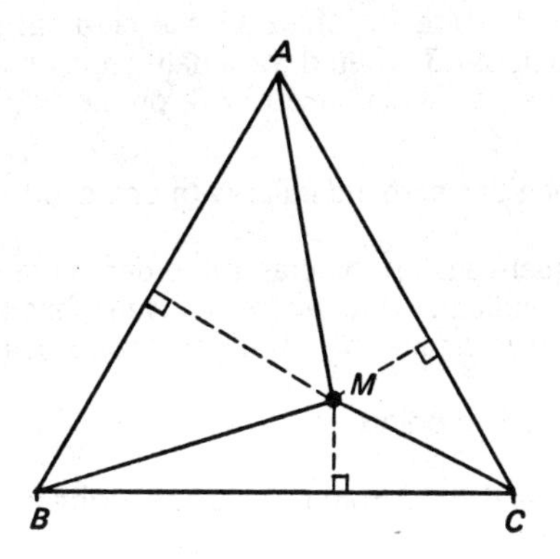

4. Construction

Draw a circle of diameter AB. From A, draw an arc of a circle of radius l. Let C be one of the intersections with the previous circle.

Join BC. From A, draw a parallel to BC. The length l of AC is the distance between the two parallels thus constructed.

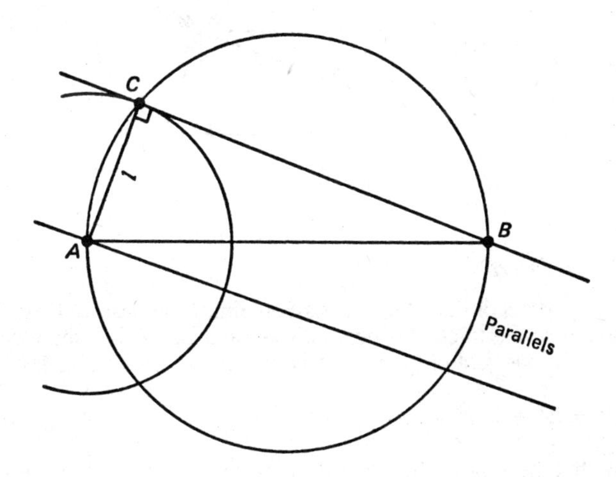

5. *Pencil, eraser, T-square, and compass*

Place the pencil on the sheet of paper. With the point of the compass, mark the ends A and B. Join them, using the T-square and pencil.

Draw the perpendicular bisector of AB by the classical process in order to define the center point of AB. Then draw a half-circle on AB. Next, using the T-square, draw the perpendicular at A to AB, on the side of the half-circle.

Place the eraser along this perpendicular with one end touching A.

Then place the T-square against the eraser in order to draw the perpendicular GM to the first perpendicular AG, where AG is the length of the eraser. The second perpendicular intersects the half-circle at two points.

Let M be one of these two points.

Next, lower the perpendicular from M onto AB, using the T-square.

Let H be the foot of that perpendicular.

Segments AH and HB provide the answer to the question. Since $AH + HB = AB$ = length of the pencil and since triangles AMH and MHB are similar, we have $AH/MH = MH/HB$. In other words, $AH \times HB = MH^2$ = length of the eraser squared.

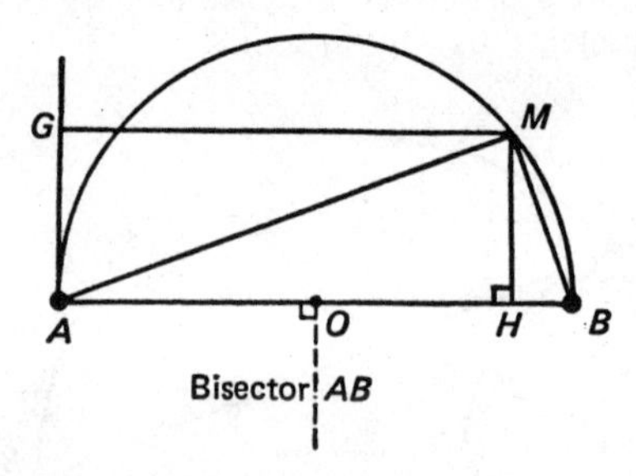

6. *Lost in the desert*

Let S be the oasis Sidi-Ben, B the oasis Ben-Sidi, I the oasis Sidi-Sidi, M the halfway point of the track between Sidi-Ben and Ben-Sidi, J the location of the Jeep, and H the foot of the perpendicular drawn from J onto IM.

Hence $JB^2 + JS^2 = 2\ JI^2$.

We know that the sum of the squares of the two sides of a triangle is equal to the sum of twice the square of the median drawn to the third side and to half of the square of that same third side. (This can easily be proved by using Pythagorean theorem.) Let us apply that theorem to triangles JHI and JHM.

We can easily deduce that

$$HI = \frac{IM^2 + BS^2/4}{2IM}$$

Therefore, HI has a constant length and the Jeep will move at right angles to IM.

The driver is correct.

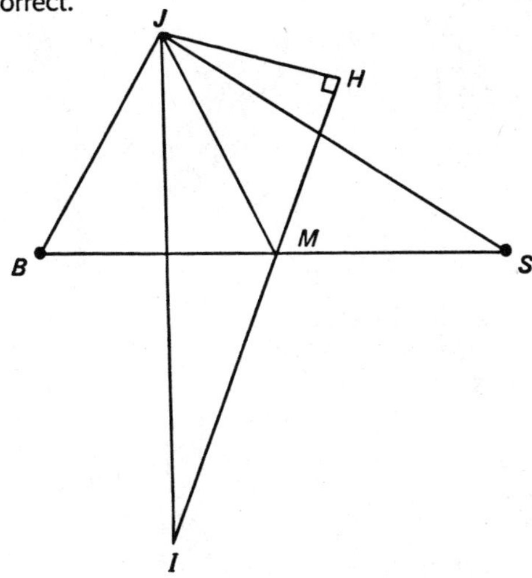

7. Cutouts

The sections of the truncated cone consist of two circles of radius 1 and 2 cm, respectively (half of 2 and 4 cm). The distance between any two corresponding points of these circles is 2 cm (4 − 2 = 2).

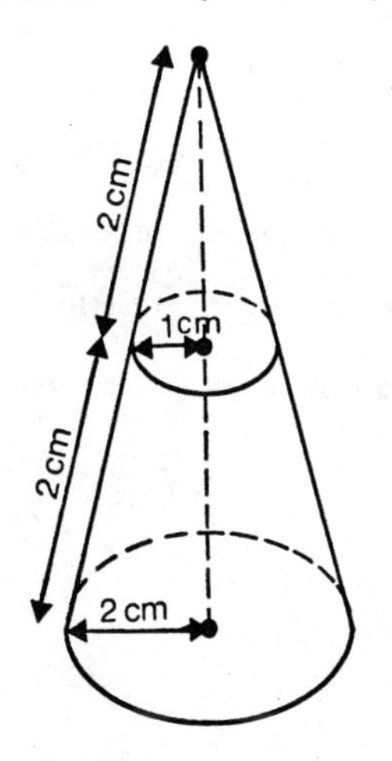

Pythagoras tells us then that the planes of these circles are $\sqrt{3}$ cm apart (because $2^2 - 1 = 3$).

The volume of the truncated cone included between the two circles is equal to the difference of the volumes of the virtual cones having each of the two circles as respective bases.

$$V = (1/3)(\pi \cdot 2^2)(2\sqrt{3}) - (1/3)(\pi \cdot 1^2)(\sqrt{3})$$
$$= 7\,\pi/\sqrt{3} = 12.697$$

The required volume is therefore 12.697 cm^3.

Note: The volume of a cone is equal to the third of the base area times the height.

8. Dublin, 1856

Let ABC be a triangle with the right angle at A. Let D, E, and F be the points of contact of the inscribed circle of center O with the three sides BC, AC, and AB, respectively.

Let $r = AE = AF$.

Let $m = CD = CE$.

Let $n = BD = BF$. Then

$$\text{Area } ABC = (1/2)(AB \cdot AC)$$
$$= (1/2)(r + m)(r + n)$$
$$= (1/2)[r^2 + r(m + n) + mn]$$

But we also have

$$\text{area } ABC = \text{area } AEOF + \text{area } FODB + \text{area } DOEC$$
$$= r^2 + 2(\text{area } BOF) + 2(\text{area } ODC)$$
$$= r^2 + nr + mr = r^2 + r(m + n)$$

Let us then substitute area ABC for $r^2 + r(m + n)$in the previous equation.

We obtain

$$\text{area } ABC = 1/2\,(ABC + mn)$$

Hence mn = area ABC

Q.E.D.

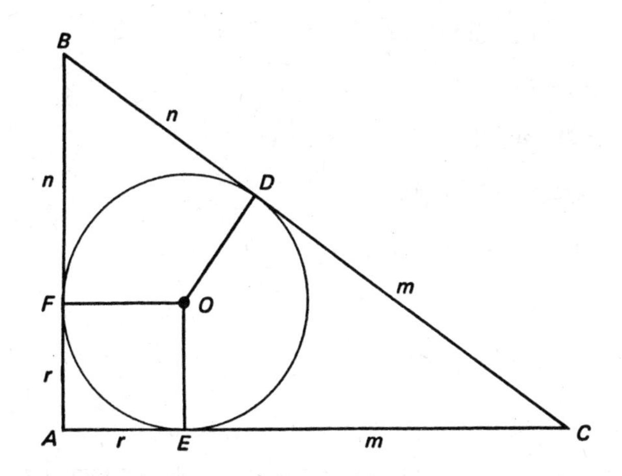

9. *Water, oil, and mercury*

The volume of a cone is equal to one-third of the product of its height times the area of its base. But the radius of the base circle for a given angle at the vertex is itself proportional to the height. The volume is therefore proportional to the cube of that height.

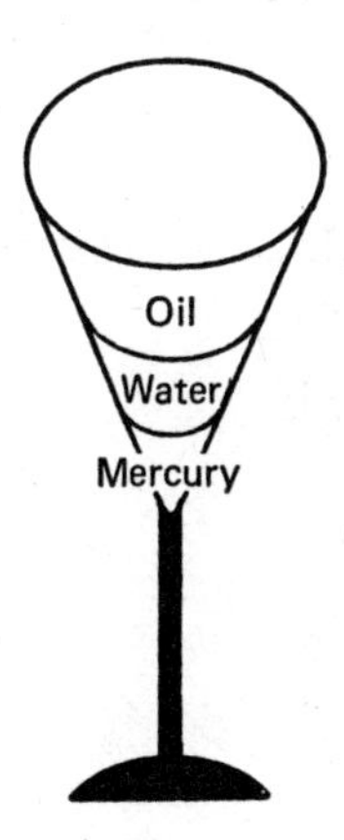

Let V_1 be the volume of the mercury, V_2 the volume of the water, and V_3 the volume of the oil. Let l be the height of each layer and let a be a coefficient of proportionality.

Let us successively consider the three cones of height l, $2l$, and $3l$:

$$V_1 = al^3$$

$$V_1 + V_2 = a(2l)^3 = 8V_1$$

$$V_1 + V_2 + V_3 = a(3l)^3 = 27V_1$$

Hence $V_2 = 7V_1$ and $V_3 = 19V_1$. Therefore:

Weight of mercury: $13.59 \cdot V_1$.

Weight of water: $1 \cdot V_2 = 7V_1$.

Weight of oil: $0.915 \cdot V_3 = 17.385V_1$.

Thus the oil weighs the most.

10. Sheik

Let BC be the side of the equilateral triangle that has the same length as its projection. Clearly, BC must be horizontal.

Let M be the center point. Let A be the third vertex of the equilateral triangle. Let H be the projection of A onto the horizontal plane that includes BC.

The angle between the planes ABC and HBC is no other than AMH, whose cosine is HM/AM. Length of AM is that of the height of the equilateral triangle:

$$AM = \sqrt{3}/2 \times \text{(length of side)}$$

Since angle BHC is a right angle and thus inscribed in a half-circle of diameter BC, MH is half the length of BC.

Therefore, the required cosine is

$$\frac{BC/2}{BC \times \sqrt{3}/2} = \frac{1}{\sqrt{3}}$$

11. Windshield wipers

Swept area: $\pi L^2 - 2S$

Area of sector OAO': $(1/6)\pi L^2$

Area of triangle OAH: $(\sqrt{3}/2)L(L/2)(1/2) = \sqrt{3}\,(L^2/8)$.

Consequently, $S = (\pi/6 - \sqrt{3}/8)L^2$.

The swept area therefore is

$$(2\,\pi/3 + \sqrt{3}/4)L^2 = 2.527L^2$$

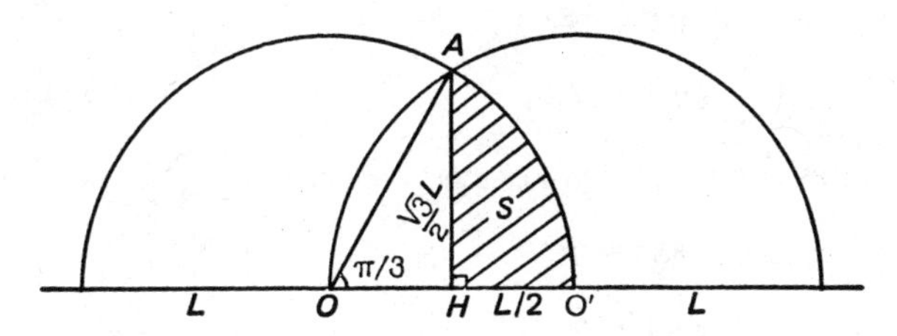

12. Figure

Let A, B, C, D, and E be the five given points. Consider the straight line AB. The three sides of the CDE triangle intersect AB in three points. The same applies for each straight line: AC, AD, and so on.

There will be as many straight lines as there are ways to choose two points among five: that is, 10. A total of $3 \times 10 = 30$ points could be generated, but each point has been counted twice. So in reality there are only 15.

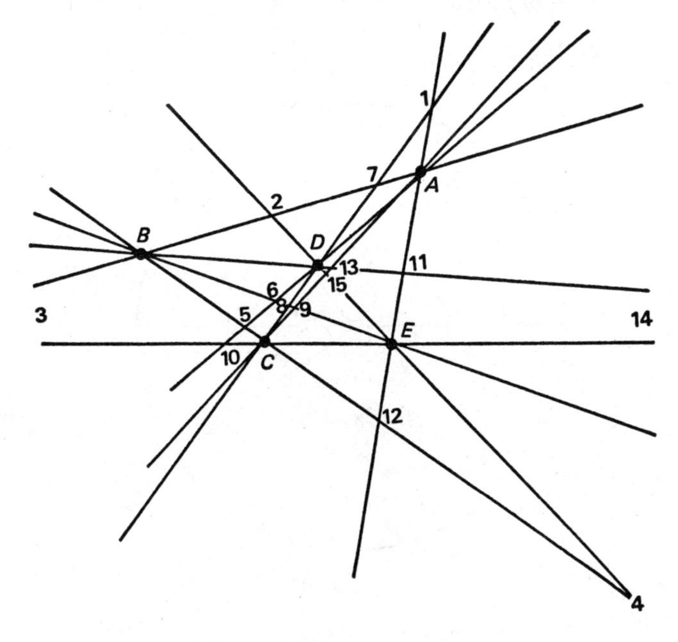

13. Christmas pie

Let O be the center of the pie. Let F be the penny. Let AB be the chord representing the second cut. Let I be the center point. Consider the triangles OAF and OBF.

We have

$$OA^2 = FA^2 + OF^2 - 2FA \cdot FI$$

$$OB^2 = FB^2 + OF^2 + 2FB \cdot FI$$

Add both sides of the equation together. That gives us

$$2OA^2 = FA^2 + FB^2 + 2OF^2 - 4FI^2 \qquad (1)$$

However, since triangle OFI is an isosceles right-angle triangle, we have

$$OF = (FI) \cdot \sqrt{2}$$

In other words

$$2OF^2 - 4FI^2 = 0$$

Hence, when this is carried into equation (1):

$$FA^2 + FB^2 = 2OA^2$$

It is therefore not sheer coincidence.

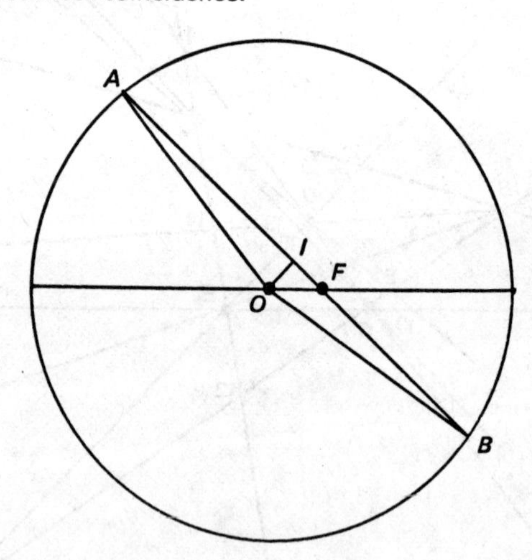

14. Revolutionary geometry

1. Pythagoras says $AC^2 = AB^2 + BC^2$

Hence

$$AC - \sqrt{AB^2 + BC^2} = 0$$

2. Area of ABC = area of *ABM* + area of *BMC, which is to say:*

$$AB \cdot BC/2 = AB \cdot d/2 + BC \cdot d/2 \quad \text{or} \quad 1/d = 1/BC + 1/AB$$

However,

$BM = d \cdot \sqrt{2}$

Hence:

$$1/AB + 1/BC - \sqrt{2}/BM = 0$$

3. If we substitute the values obtained above in the expression for *E, we get*

$E = 1789$ (which is the date of the French Revolution).

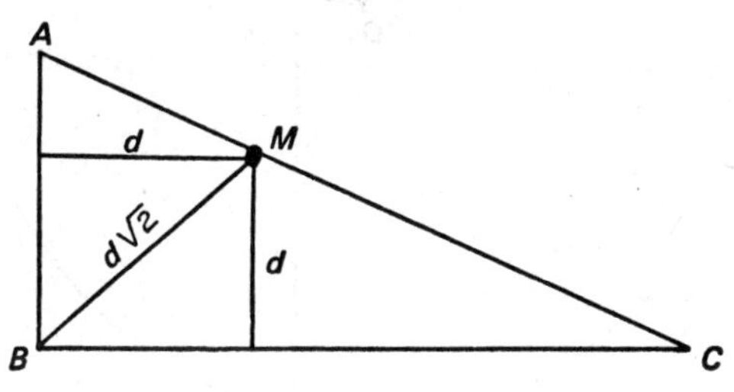

15. High mass

Let *AB* and *CD* be the two segments of the cross.

Let *M* be its center point (the intersection of *AB* and *CD*).

Let *O* be the center of the window and *H* its projection on *AB*.

Let α be the angle $\widehat{OMH}$.

Consider the right triangle *OHB*:

$$HB^2 = OB^2 - OH^2 = OB^2 - OM^2 \sin^2 \alpha^2$$

AB, however, is twice the length of *HB. Hence*

$$AB^2 = 4OB^2 - 4OM^2 \sin^2 \alpha^* = (4 - \sin^2 \alpha) \quad m^2$$

*Square meters

To calculate CD^2 we will proceed in a similar fashion, the new angle that plays the role of α is, of course, equal to $90° - \alpha$. We therefore find that

$$CD^2 = 4\,OC^2 - 4OM^2\sin^2(90° - \alpha) = (4 - \cos^2\alpha) \qquad m^2$$

Hence

$$AB^2 + CD^2 = 7 \text{ m}^2$$

Our daydreaming parishioner arrives at the figure seven, which is the number of capital sins.

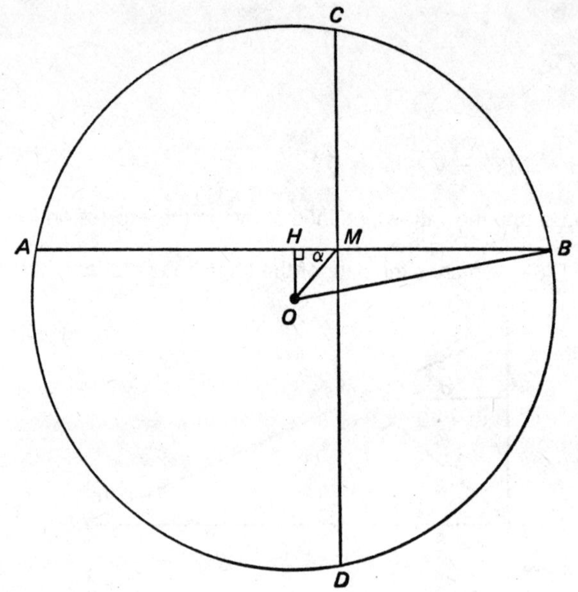

$$AB^2 + CD^2 = \text{number of capital sins}$$

16. Legacy

Let H be the foot of the perpendicular drawn from O onto AB and K that of the perpendicular drawn from O onto CD.

The area left to Peter is

$$(1/2)(OH \cdot AB) + (1/2)(OK \cdot CD) = AB \cdot HK/2$$

The total area of the parallelogram-shaped field $ABCD$ is $AB \cdot HK$.

Peter now owns half of this area, John the other: Neither son has been favored.

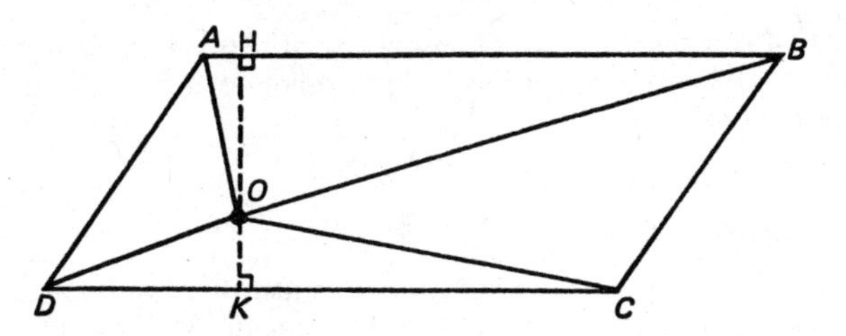

17. Hexagon

1. With six different points one can construct as many straight lines as there are ways to choose two items among six, which is

$$C_6^2 = \frac{6!}{2!\,4!} = 15$$

Among these 15 straight lines, 6 are sides, which leaves 9 diagonals.

2. Fifteen straight lines generally have as many points of intersection as there are different ways to choose 2 items among 15, thus:

$$C_{15}^2 = \frac{15!}{2!\,13!} = 105$$

But in this case, each of the six vertices of the hexagon corresponds to the intersection of five different straight lines. Each vertex therefore represents the congruence of 10 vertices because

$$C_5^2 = \frac{5!}{3!\;2!} = 10$$

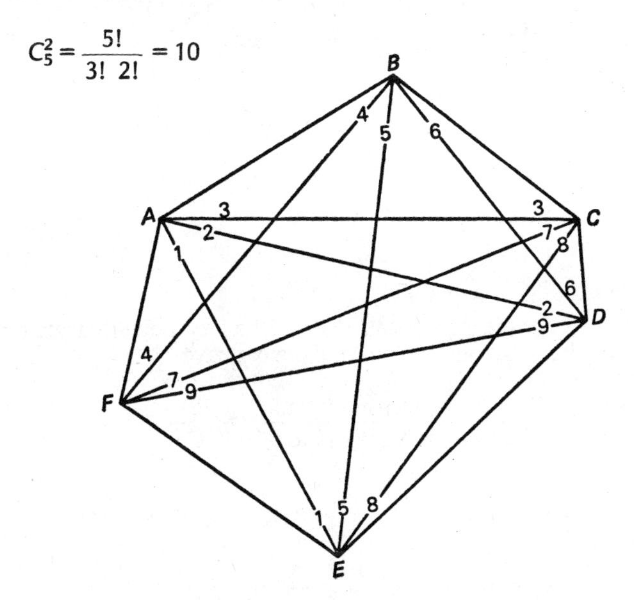

The six actual vertices therefore represent a total of 60 virtual vertices. Hence there remain $105 - 60 = 45$ intersection points that are distinct from the vertices of the hexagon.

18. *Intersections*

Let $ABCD$ be a quadrilateral. Let M, N, P, and Q be the respective center points of sides AB, BC, CD, and DA. In the triangle ACB, the segment MN, which joins the center points of sides AB and BC, is parallel to AC and equal to half its length. The same can be said of QP. The quadrilateral $MNPQ$, which has two equal and parallel sides, is therefore a parallelogram. Hence the diagonals MP and QN intercept at their center points.

Consider next the center points of AC and BD (referred to as E and F, respectively). In triangle ABD, it can be seen that FQ is equal to $AB/2$ and parallel to it. The same can be said of NE in triangle ABC. Thus $NEQF$ is a parallelogram and its diagonals FE and NQ intersect at their center points.

MP, QN, and FE therefore meet at their center points.

Q.E.D.

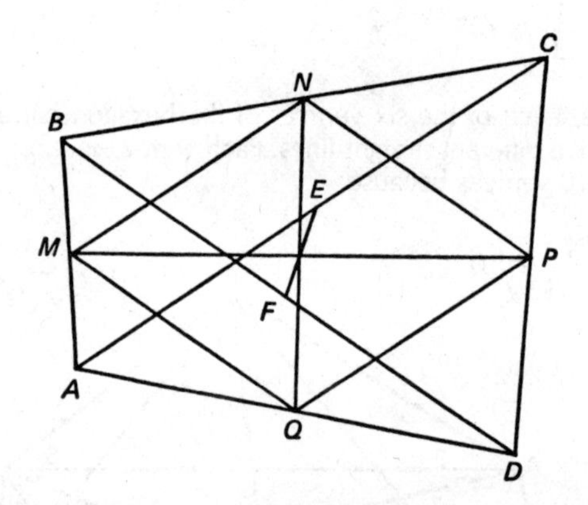

19. *Italy, 1678*

We will agree from the start to call $S(\cdot, \cdot, \cdot)$ the area of the triangle defined by the three letters in parentheses. We therefore have

$$\frac{S(BPX)}{S(CPX)} = \frac{(1/2)BX \cdot \text{distance from } P \text{ to } BC}{(1/2)CX \cdot \text{distance from } P \text{ to } BC} = \frac{BX}{CX}$$

Similarly, we have

$$\frac{S(BAX)}{S(CAX)} = \frac{BX}{CX}$$

Therefore,

$$\frac{S(BAX) - S(BPX)}{S(CAX) - S(CPX)} = \frac{BX}{CX} \quad \text{or} \quad \frac{S(ABP)}{S(ACP)} = \frac{BX}{CX}$$

In an analogous way we can show that

$$\frac{S(BPC)}{S(BPA)} = \frac{CY}{AY} \quad \text{and} \quad \frac{S(ACP)}{S(BPC)} = \frac{AZ}{BZ}$$

The product of the ratios

$$\frac{BX}{CX} \cdot \frac{CY}{AY} \cdot \frac{AZ}{BZ}$$

therefore equals

$$\frac{S(BAP)}{S(ACP)} \cdot \frac{S(BPC)}{S(BPA)} \cdot \frac{S(APC)}{S(BPC)} = 1$$

Q.E.D.

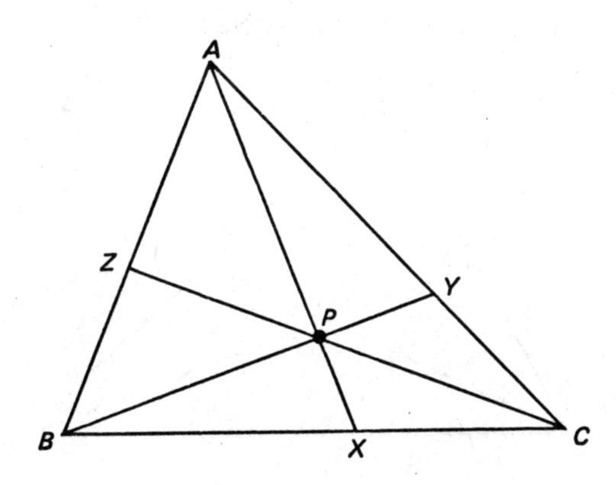

20. Medians

Let ABC be any triangle and G the intersection point of the three medians. We know that G is located two-thirds of the way along each median from the corresponding vertex of the triangle.

Consider segment AB. Since a straight line represents the shortest distance between two points, we have

$$AB < AG + GB$$

Similarly, we can establish the inequalities

$$BC < BG + GC \qquad \text{and} \qquad CA < CG + AG$$

Let us add together the corresponding sides of these three inequalities:

$$\text{perimeter} < 2\,(AG + BG + CG)$$

or

perimeter < 2(2/3 of the median drawn from A + 2/3 of the median drawn from B + 2/3 of the median from C

Thus we have

$$\text{sum of the lengths of the medians} > (3/4)(\text{perimeter})$$

The answer to the question posed is therefore "no."

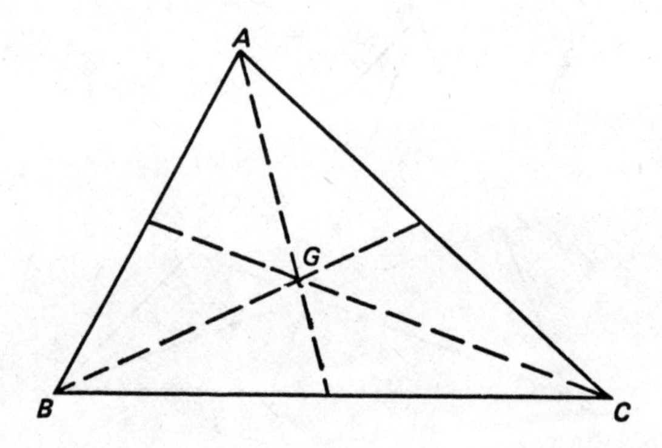

21. Parallelogram

Angle $\widehat{DAB}$ being constant, vertex A of the parallelogram will remain on a particular circle passing through M and N when the parallelogram moves. Let P be the point of intersection of this circle with diagonal AC. Angle $\widehat{DAC}$ ($\widehat{NAP}$) being constant, point P will not vary during the motion of the parallelogram. It is the fixed point we are looking for. *Note*: This theorem was formulated by Maurice d'Ocagne in 1880 while he was studying at the École Polytechnique in Paris. It corresponds to the following theorem of statics (Abel Transon, 1863): "When two force vectors are rotated through the same angle and in the same direction around their respective application points, the resultant force rotates through the same angle and passes through a fixed point."

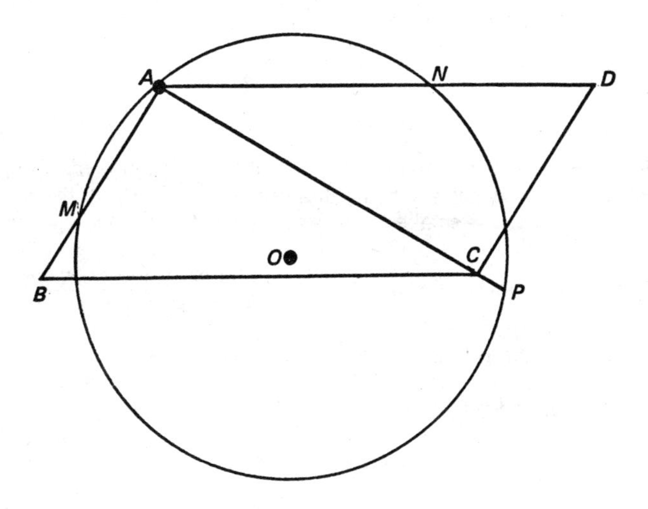

22. Lighthouse

Let S be the top of the lighthouse. Let A be a point on the horizon.

Let O be the center of the earth. Triangle OAS is a right-angle triangle since SA is tangent to the earth, which we consider to be a perfect sphere.

We thus have $OS^2 = OA^2 + AS^2$

but $OA = \dfrac{40{,}000{,}000}{2\pi}$ (in meters)

and $OS = OA + 125.7 = OA + 40\pi/OS^2 \simeq OA^2 + 2 \cdot OA = 40\pi$

Consequently,

$$AS = \sqrt{OS^2 - OA^2} \simeq \sqrt{2 \cdot \frac{40\pi \cdot 40{,}000{,}000}{2\pi}} = 40{,}000$$

The horizon is therefore approximately 40 km from the top of the lighthouse.

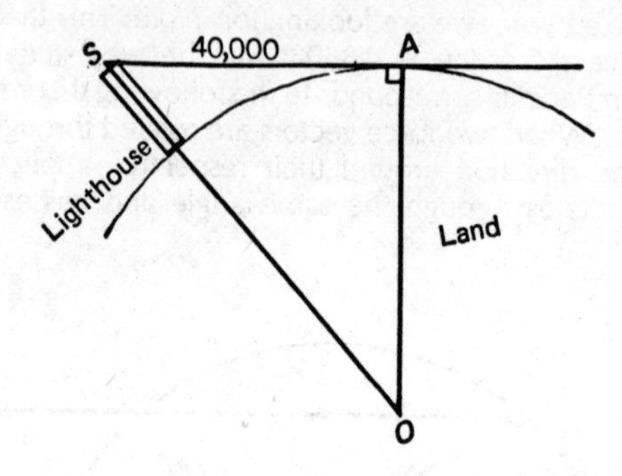

23. Polygons

If p is a divisor of n, the number of sides of the polygon is n/p. If p is not a divisor of n, it is necessary to multiply the ratio n/p by x, the number of complete rotations: x = smallest number y such that yn/p is a whole number:

$$\text{number of sides} = n(x/p) = \frac{\text{smallest common multiple of } n \text{ and } p}{p}$$

When $n = 10$, and $p = 2$; number of sides: $10/2 = 5$.

When $n = 10$, and $p = 3$; number of sides: $30/3 = 10$.

When $n = 10$, and $p = 4$; number of sides: $20/4 = 5$.

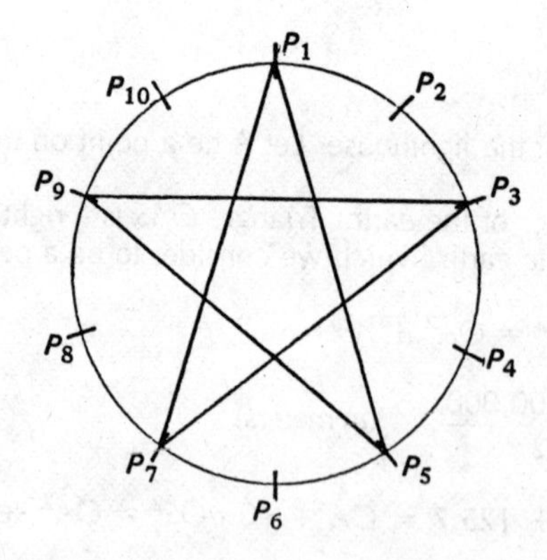

24. *Walking the dog*

Let O and O' be the respective positions of Mr. and Mrs. Jones at a given time.

If the dog was connected only to Mr. Jones, the area he could cover would be π (area of a circle with a radius of 1 m^2).

The same would be true if he were attached only to Mrs. Jones. In reality, however, he can only move over a common area of the two circles with radii of 1 meter and whose centers are 1 meter apart.

Let A and B be the points of intersection of these circles (centers O and O', radius 1 meter).

Area of the diamond $OAO'B$: 2(area OAO') = $\sqrt{3}/2$.

Area of the sector OAB of the circle centered on O: $\pi/3$.

Area of the sector $O'AB$ of the circle centered on O': $\pi/3$.

The area of intersection of the two circles is therefore

$$\pi/3 + \pi/3 - \sqrt{3}/2 = 2.094 - 0.866 = 1.228$$

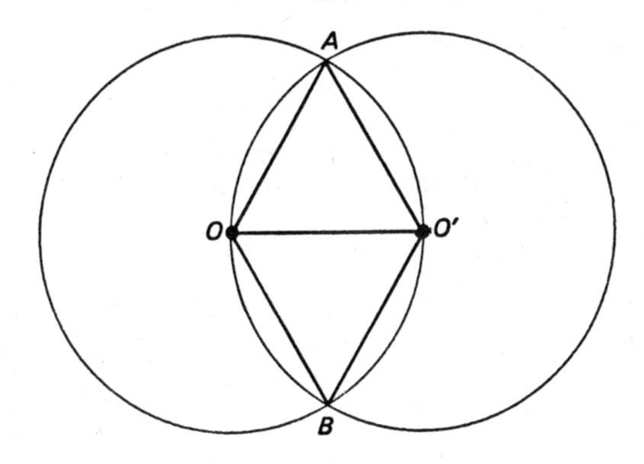

25. *Quadrilateral*

Let $ABCD$ be this quadrilateral and AC a diameter. Let H and K be the projections of A and of C on BD, respectively. Let E be the point of intersection of CK with the circle.

Angles $\widehat{ADB}$ and $\widehat{ECD}$ are equal because their respective sides are perpendicular to one another. The same is therefore true of arcs AB and ED and, consequently, of chords AB and ED. Arcs AD and EB are also equal (they can be obtained by subtracting from each of the previous arcs the same arc AE).

Therefore, angles $\widehat{ABD}$ and $\widehat{BDE}$ are also equal. Consequently, the right-angle triangles ABH and EKD are equal, as are segments BK and DH. The projections of the opposite sides of this quadrilateral on the diagonal which is not a diameter, are therefore equal.

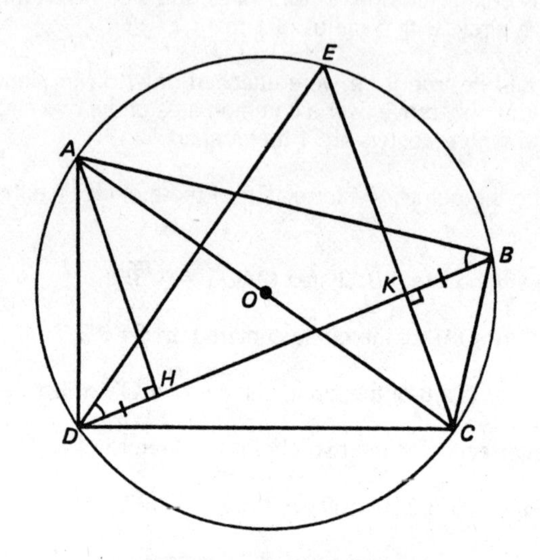

26. The ramparts

Let us call O the center of the half-circle and A and B the intersection points of the tangents at P_1 and P_3 and at P_2 and P_3, respectively.

Angles $\widehat{P_1AP_3}$ and $\widehat{P_2BP_3}$ are obviously supplementary. Their half-angles ($\widehat{OAP_1}$ and OBP_2) are therefore complementary. As a result, the right-angle triangles OAP_1 and OBP_2 are similar. Therefore,

$$AP_1/OP_1 = OP_2/BP_2 = OP_1/BP_2$$

or

$$AP_1 \times BP_2 = OP_1^2/2AP_1 \times 2BP_2 = 4OP_1^2$$

hence

$$(P_1A + AP_3) \times (P_3B + BP_2) = 4 \times OP_1^2 = \widehat{P_1P_2^2}$$

Q.E.D.

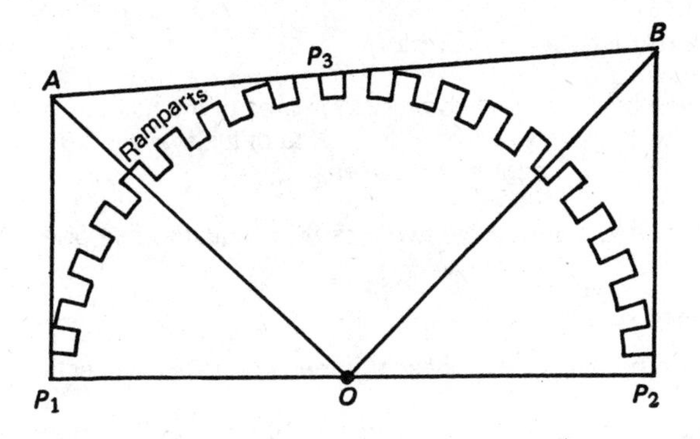

27. Equal areas

Let a be the length of AB or AC.

Area of the triangle ABC: $a^2/2$.

Length of the hypotenuse $BC = a\sqrt{2}$.

Area of the half-circle of diameter BC:

$$\frac{1}{2}\pi\frac{BC^2}{4} = \pi\frac{a^2}{4}$$

Let O be the center of the circle tangent at B to AB and at C to AC. OBC is an isosceles right-angle triangle equal to ABC: area $OBC = a^2/2$.

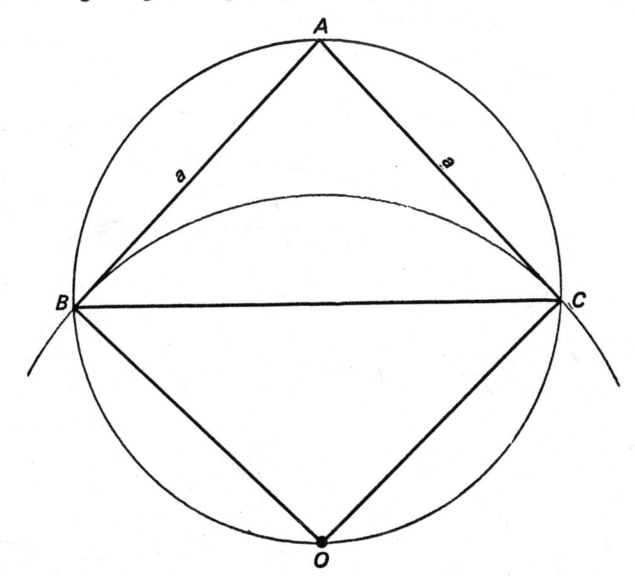

Area of the quarter-circle OBC: $\pi a^2/4$.

Area included between the two arcs BC = area of the half-circle of diameter BC – (area of the quarter-circle OBC – area of the triangle OBC) = $\pi a^2/4$ – $(\pi a^2/4 - a^2/2) = a^2/2$ = area ABC.

The area included between the two arcs BC is equal to that of ABC.

28. *Tangency*

Let O and O' be the centers of each of the two circles, respectively.

Let P be the point where they are tangent. OO' passes through P.

Let AA' and BB' be the two straight lines drawn through P, A and B being on the circle of center O, and A' and B' being on the circle of center O'.

Angles $\widehat{APO}$ and $\widehat{O'PA'}$ are obviously equal. Since the triangles AOP and $A'O'P'$ are isosceles, angles $\widehat{AOP}$ and $\widehat{A'O'P'}$ are also equal. In the circle of center O, angle $\widehat{ABP}$, which intercepts arc AP, is equal to half of the corresponding angle at the center $\widehat{AOP}$. Similarly, $\widehat{PB'A'}$ is equal to half of $\widehat{PO'A'}$. Angles $\widehat{PBA}$ and $\widehat{PB'A'}$ are therefore equal. AB and $A'B'$ are thus parallel. The quadrilateral $ABA'B'$ is a trapezium.

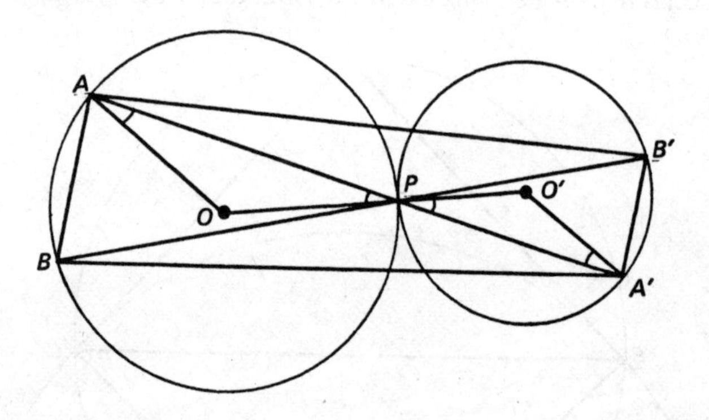

29. *Triangle*

Construct an angle equal to the given angle. Let A be its vertex. Let D and E be the points located on its two sides at a distance of A equal to the half-perimeter given. Let O be the point where the perpendiculars from D and E meet. Draw a circle of center O, with radius OD or OE. Then draw a second circle of center A so that its radius is equal to the given height. Draw the inside tangent common to the two circles. Let B and C be the points where the tangent meets the sides of angle $\widehat{A}$. We have $AB + BC + CA = AD + AE =$ perimeter. Triangle ABC is the triangle sought.

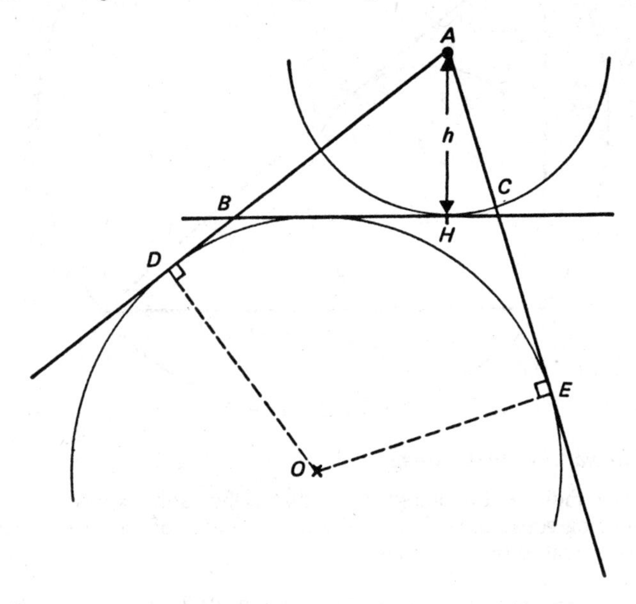

30. *Right-angle triangle*

Let ABC be a right-angle triangle (right angle at A).

Let D, E, and F be the points of contact of the inscribed circle (D being on BC, E on CA, and F on AB).

$AF = AE = r$ (radius of the inscribed circle).

$CE = CD$.

$BF = BD$.

Thus $CE + BF = CD + DB = CB = 2R$ (twice the radius of the circumscribed circle)

Hence $AB + AC = AF + FB + AE + EC = 2r + 2R$.

Q.E.D.

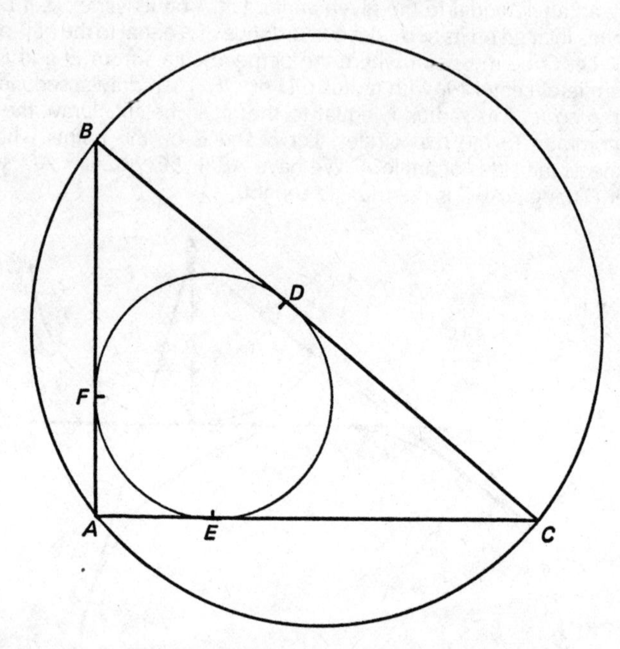

31. If the earth were an orange

Let r be the radius of the orange and R that of the earth, expressed in meters. The red string goes from a length of $2\pi r$ to a length of $2\pi(r + 1)$; that is, its length is increased by 6.283 meters.

The blue string goes from a length of $2\pi R$ to $2\pi(R + 1)$, increasing by 6.283 meters as well.

The two increases are identical.

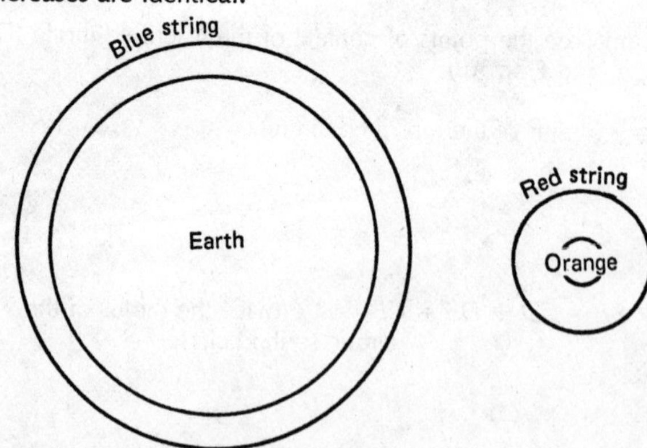

32. *Siesta*

Albert has placed the tips of his middle and index fingers on the terrace (a horizontal surface), maintaining a constant angle of 20° between his two fingers. When his hand rotates around these two points of contact, the angle made by the shadows cast by his fingers changes. It is 20° when his hand is flat and 180° when his hand is in a plane parallel to the rays of the sun. Since this change in angle is obviously continuous, there will be a time when the two shadows will make an angle of exactly 90° (a value somewhere between 20° and 180°). This position is therefore easy to find. Like Albert, we'd soon be fast asleep.

33. *Statues*

Let P_1, P_2, P_3, and P_4 be four successive bridges.

Distance from P_1 to Vercingetorix: $P_1P_2/2$.

Distance from P_1 to Henri IV: $P_1P_2 + P_2P_3 + P_3P_4/2$.

Thus the distance from P_1 to the midpoint between the two statues is:

$$(1/2)(P_1P_2/2 + P_1P_2 + P_2P_3 + P_3P_4/2)$$

Distance from P_1 to Charlemagne: $P_1P_2 + P_2P_3/2$.

Distance from P_1 to Louis XIV: $(P_1P_2 + P_2P_3 + P_3P_4)/2$.

Hence the distance from P_1 to the midpoint between the two statues is

$$(1/2)(P_1P_2 + P_2P_3/2 + P_1P_2/2 + P_2P_3/2 + P_3P_4/2)$$

The two locations proposed for Napoleon's statue therefore coincide.

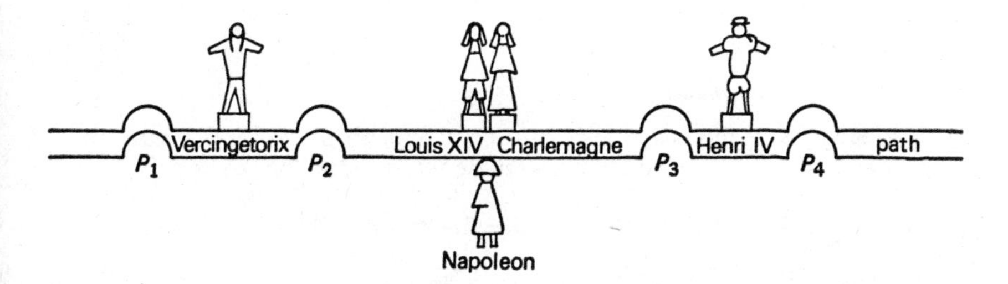

34. Street talk

Let S be the school, T the town hall, P the Protestant church, and C the Catholic church. The area of triangle TCP is equal to the sum of the areas of triangles TCS and TPS. This can be written as follows:

$$500 \cdot TC/2 + 500 \cdot TP/2 = TP \cdot TC/2$$

As a result,

$$TC + TP = TP \cdot TC/500$$

In other words,

$$1/TC + 1/TP = 1/500 = 0.002$$

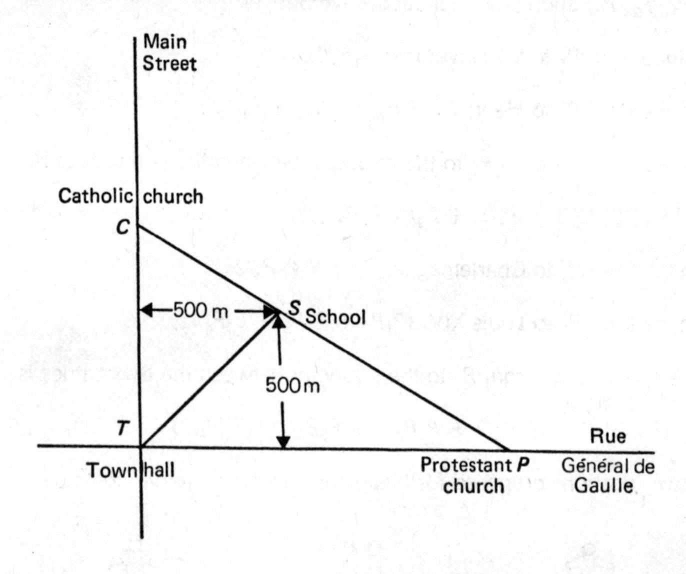

35. View from a plane

Three-fourths of the island accounts for one-fourth of the view. Thus, the whole island = 1/3 of the scenery and the sea = 2/3 of the scenery.

The sea not hidden by the cloud: $1/2 - 1/4 = 1/4$ of the view.

Thus the proportion of sea hidden by the cloud is

$$\frac{2/3 - 1/4}{2/3} = 5/8 = 0.625$$

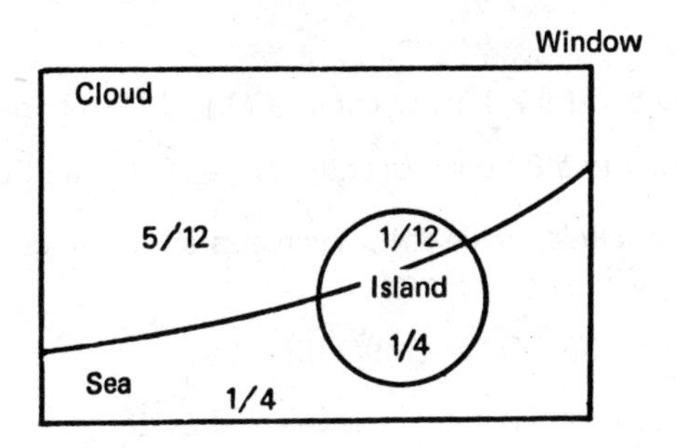

36. Pedestrian zone

Consider a square matrix composed of seven lines and seven columns (one for each street): Fourth Street, for example, corresponding to the fourth line and the fourth column. Put a 1 in a square if the corresponding streets cross and an O if they don't cross (the diagonal will be all Os).

First observation: If there is a 1 at the intersection of the *i*th line and the *j*th column, there will also be a 1 at the intersection of the *j*th line and the *i*th column. The matrix is therefore symmetrical around the first diagonal and has an even number of 1s.

Second observation: If each of the seven streets cut three others, no more, no less, we should find three 1s in each line and the total number of 1s should be $7 \times 3 = 21$, which is an odd number. This is impossible.

The answer to the question posed is therefore "no."

		j		*i*			
	0						
j		0		1			
			0				
i		1		0			
					0		
						0	
							0

37. Two circles

Let O be the center of the smaller circle. CO and DA are parallel (they are both perpendicular to DB); consequently, angles $\widehat{DAB}$ and $\widehat{COB}$ are equal.

But in the smaller circle, angle $\widehat{CAO}$ intercepts an arc whose corresponding center angle is $\widehat{COB}$.

Therefore, $\widehat{CAO} = (1/2)\widehat{COB} = (1/2)\widehat{DAB}$.

Hence $\widehat{CAO} = \widehat{DAC}$. AC is thus the bisector of $\widehat{DAB}$.

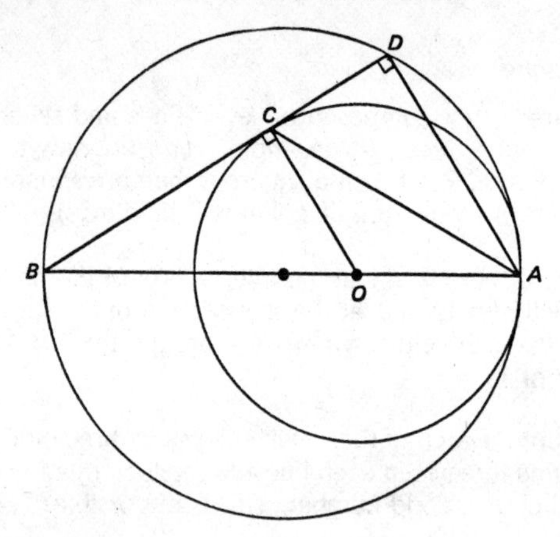

38. Crescents

Area ABC = area of the half-circle of diameter BC − area of circle segment AB − area of circle segment $AC = \pi BC^2/8$ − areas of the two circle segments.

Area of the outside crescent built on AB = area of the half-circle of diameter AB − area of circle segment $AB = \pi AB^2/8$ − area of circle segment AB.

Area of the outside crescent built on AC = area of the half-circle of diameter AC − area of circle segment $AC = \pi AC^2/8$ − area of circle segment AC.

Pythagoras tells us that $BC^2 = AB^2 + AC^2$

Consequently, $\pi BC^2/8 = \pi AB^2/8 + \pi AC^2/8$

As a result: area of the half-circle of diameter BC = area of the half-circle of diameter AB + area of the half-circle of diameter AC.

Subtract from each side of this equation the sum of the areas of the two circle segments. We conclude: The sum of the areas of the two small crescents equals that of triangle ABC.

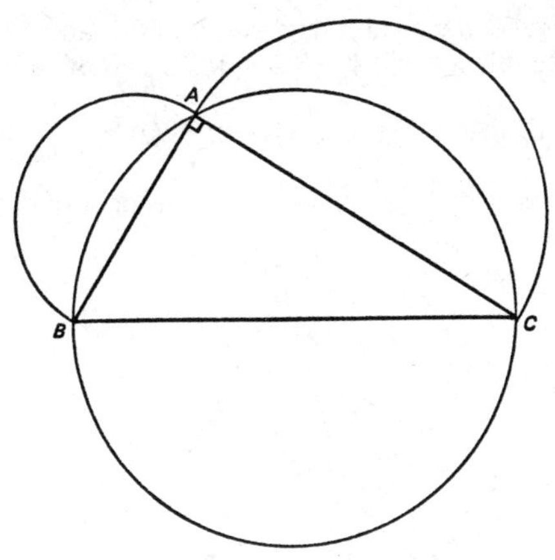

39. *The two bridges*

Let A and B be the two towns. Let B' be such a point that BB' is a line equal and parallel to each future bridge, B' being closer to the canal than B.

Let D be the straight line representing the canal bank closest to A.

The intersection of the bisector of AB' with D will give the starting point of the first bridge, while the intersection of AB' itself with D will give that of the second bridge (the proof is trivial).

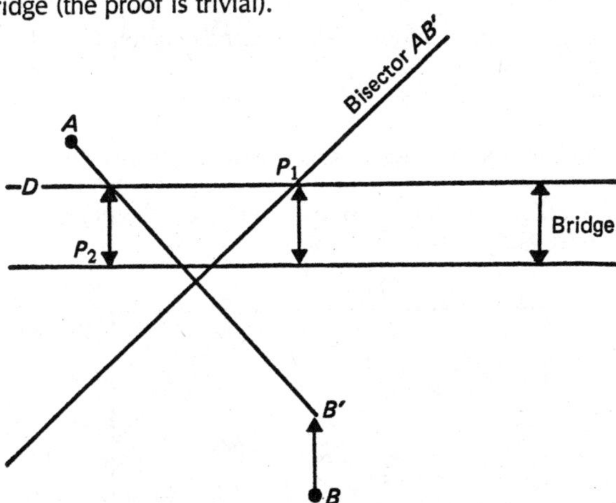

40. *Three perpendiculars*

Let ABC be an equilateral triangle and P any point inside. Let H, K, and L be the three feet of the perpendiculars drawn on AB, BC, and CA, respectively. Let h, k, and l be the respective lengths of PH, PK, and PL.

The sum of the three surfaces of triangles APB, BPC, and CPA equals that of ABC. This gives the relationship $h \cdot AB/2 + k \cdot BC/2 + l \cdot CA/2 =$ area ABC

Now $AB = BC = CA$. As a result, $h + k + l = (2/AB) \times$ area of ABC

The sum of the three lengths is therefore independent of the actual position of point P.

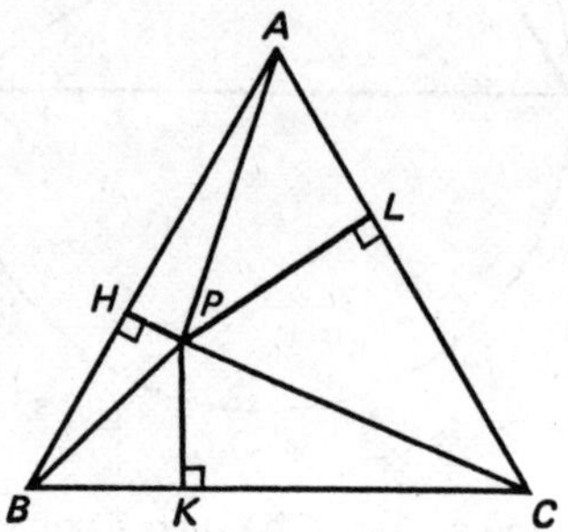

41. The four suits of armor

The section of each niche reserved for armor will be an isosceles right-angle triangle of side x. The hypotenuse will therefore be : $\sqrt{2}x$. This hypotenuse is equal to any of the eight equal sides. It will therefore be necessary to divide each of the sides of the initial square into three parts:

x for the left-hand niche. $\sqrt{2}x$ for the side of the octagon.

x for the right-hand niche.

Thus $(2 + \sqrt{2})\, x = 3.414$ meters. Consequently, $x = 1$ meter.

The sides of the triangular section of each niche are therefore
1 meter, 1 meter, and 1.414 meters

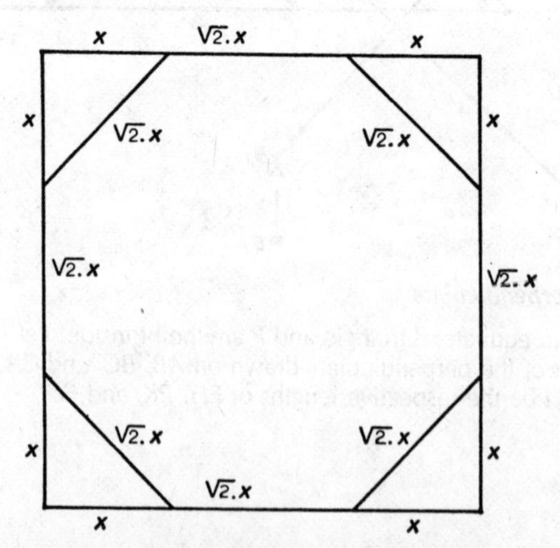

5. The wonderful world of whole numbers

1. Barking

A trained dog can count in a numeration system in base 4. The four symbols he uses are O for zero, F for 1, R for 2, and W for 3. What number is he trying to give when he barks "ROWFROWF"?

2. The demonstration

If we set out by ranks of 10, we will be one short. We will also be one short if we set out by ranks of 9, 8, 7, 6, 5, 4, 3, and even 2. Yet there are fewer than 5000 participants. How many are we?

3. Year of birth

Take away from your year of birth the sum of the four numerals that make it up. You will end up with a number divisible by 9. Why is this?

4. Arthmetic

What are the eight numerals represented by the eight letters E, G, H, I, N, T, W, Y in the following sum?

```
  TWENTY
+ TWENTY
+ TWENTY
+    TEN
+    TEN
--------
= EIGHTY
```

5. Historical arithmetic

A historian found one day that the year of the discovery of the West Indies by Christopher Columbus and the year of the election of Calvin Coolidge as president were formed by the same numerals written in a different order. In addition, the remainders, after dividing each of the two dates by 9, were the same. He then looked for other pairs of historical dates that had the first characteristics without the second. He failed to find any. Help him.

6. Spanish arithmetic

$4 + 4 + 4 + 4 + 4 = 20$

In Spanish, this can be written:

```
    CUATRO
+   CUATRO
+   CUATRO
+   CUATRO
+   CUATRO
----------
=   VEINTE
```

How are the 10 numerals represented by the 10 letters A, C, E, I, N, O, R, T, U, V for this sum to come out right?

7. Graduation procession

The principal of a certain high school noted with amusement that she could determine the number of pupils at the school by multiplying the difference of the squares of the number of physical education teachers and of the number of Russian teachers by the product of these last numbers. Given this information, do you think it is possible to organize the graduation ceremony with all the pupils parading in full ranks of three?

8. Anniversary

To celebrate their wedding anniversary, Mr. Smith takes his wife out to lunch at a fancy restaurant. As he leaves, he discovers that he only has one-fifth of his money left and that what he has left in cents represents what he originally had in dollars, whereas he had five times fewer dollars than he had cents to start with. How much was the check?

9. The bus

The number of passengers on a bus stayed constant from the start until the second stop. After that, it was discovered that as many people got on at each stop as had gotten on at the two previous stops, whereas as many people got off at each stop as had gotten off at the preceding stop. At the tenth stop, the end of the line, 55 passengers got off. How many were on the bus between the seventh and eighth stops?

10. A memorable birthday

Do you remember Lucy's fortieth birthday? It was December 28, 19—. Since then, Lucy has grown older. I noticed that half her age was equal to twice the sum of the digits that make up her age. Fill in the missing digits of the year of her fortieth birthday.

11. A story of 1s and 2s

Write, side by side, the numeral 1 an even number of times. Take away from the number thus formed the number obtained by writing, side by side, a series of 2s half the length of the first number. You will always get a perfect square. For instance,

$$1111 - 22 = 1089 = (33)^2$$

Can you say why this is?

12. Great-grandfather's family

I am 91 years old. I have 4 children, 11 grandchildren, and many great-grandchildren. If you ask me how many I have exactly, I will answer only that the product of their number by the number of my grandchildren and by my age can be written as a number made up by writing side by side this unknown number, a zero, and again the unknown number. What can you deduce from all this?

13. Caramels

Melissa hands out 26 caramels to her four smaller brothers. They each eat several and after an hour she discovers that each has the same number left. Knowing that the eldest ate as many as the third, that the second ate half his initial share, and that the fourth ate as much as the other three put together, how did Melissa divide the caramels?

14. Square

Can the square of a whole number end in three identical digits other than zeros?

15. Martin and Mildred

One day, Martin said to Mildred: "I am three times older than you were when I was as old as you are now." Mildred answered: "When

I am as old as you are now, the sum of our ages will be 77." How old is Martin? How old is Mildred?

16. How many children were you?

If you ask me such a question, I will reply that my mother dreamed of having at least 19 children but her dream did not come true, that my sisters were three times more numerous than my first cousins, and that I had half as many brothers as I had sisters.

17. Airline company

An American airline connects a number of large towns together. Each pair of towns is connected by a direct flight. Next year, the company intends to add 76 new flights that would incorporate a certain number of new towns into the existing network, with direct connections to all other towns. How many towns are served at the moment, and how many will be serviced next year?

18. Latin exam

The Latin exam was scored from 0 to 20 using whole numbers only. Michael's mark was better than 10%. Claude's was less. What mark did each get, given that if one takes away from each of these scores the third of the smallest score, the remainder of the larger number of the two will be three times the remainder of the smaller?

19. Counting by fingers

If you have trouble remembering your times-9 multiplication table, try the following: To find the value of $9 \times n$ (n being any number chosen at random), lay your two hands flat on the table. Lift the *n*th finger starting from the left. The answer has for digits representing the "tens" the number of fingers still on the table to the left, and for the digit representing the "units" the number of fingers still on the table to the right. Try it.

Now that you have checked out the trick, can you explain it?

20. Cluck-cluck

A chicken could count according to a numbering system in base 4. The four symbols that it used were U, K, L, and C. What were the numerical values assigned to each of these letters given that when it wanted to say 181,425, it went "cluck-cluck"?

21. Horse race

On a fine spring afternoon, I went to the races at Bowie. I bet on a first horse and doubled the money I had with me. Encouraged by this success, I bet $60 on a second horse and lost it. Thanks to a third horse, I doubled my assets. But a fourth horse caused me to lose $60 again. A fifth horse enabled me again to double the amount of money in my possession. A sixth horse on which I bet $60 was disastrous: I had no money left. How much money did I have to start with?

22. Age difference

The numerals of the years of birth of John and Jack add up to the same figure. Knowing that their ages both start with the same numeral, can you compute their age difference?

23. Division

A division is set. If one increases the dividend by 65 and the divisor by 5, one finds that the quotient and the remainder do not change. What is the quotient?

24. Number switching

A three-digit number increases by 45 if the two right-hand numerals are interchanged and decreases by 270 if the two left-hand numerals are interchanged. What would happen if one interchanged the left-hand numeral with the right-hand numeral?

25. School fire

The different classrooms in a school each have the same number of pupils. As the result of a fire, six of the classrooms were completely destroyed. It was necessary to redistribute the pupils by adding five pupils to each of the remaining classrooms. It was then realized that 10 of the other classrooms were unusable because of water damage. Hence a new redistribution of pupils was required and 15 more pupils were added to the usable classrooms. How many pupils are there at the school?

26. Brother and sister

"Sister, you have as many brothers as you have sisters." "Brother, you have twice as many sisters as you have brothers." Can you deduce from this conversation how many children there are in the family?

27. Large families

The Martin family has more children than the Davis family. Knowing that the difference of the squares of the number of children in each family is 24 and that both families have more than one child, how many children are there in the Martin family?

28. Eleanor's riddle

When I saw Eleanor, I thought she was beautiful. After a brief conversation, I told her how old I was and asked her own age. She answered: "When you were as old as I am now, you were three times older than I was. When I am three times older than I am now, our ages will add up to 100." I couldn't understand this language and told her so. This was not to her liking and she walked away. How old was she?

29. Ursula and the cats

If you ask old Ursula how many cats she has at home, she answers sadly: "Four-fifths of my cats plus four-fifths of a cat." How many cats does this add up to?

30. New math

My son learned how to count in a base different from 10, so that, for instance, instead of writing 136, he writes 253. In what base does he count?

31. Multiplication

Can you complete the following multiplication?

```
    ....
×     .9
--------
   .7547
.....
--------
25886.
```

32. Chinese numbers

Think of a number between 1 and 26. Look at the following table of six squares, one square at a time.

1 4 7 10 13 16 19 22 25	2 5 8 11 14 17 20 23 26	3 4 5 12 13 14 21 22 23
6 7 8 15 16 17 24 25 26	9 10 11 12 13 14 15 16 17	18 19 20 21 22 23 24 25 26

Each time the number you have picked belongs to one of the squares, write down the number in the top left-hand corner. Add together all these numbers.

For example, 16 is in the first, fourth, and fifth squares. If the first numbers of each of these squares are added together, we get 1 + 6 + 9 = 16, which is the original number we picked. Can you explain this?

33. The smallest possible

Which is the smallest number which, when divided by 2, 3, 4, 5, and 6, will give 1, 2, 3, 4, and 5 as remainders, respectively?

34. For those under 16

Tell me in which columns your age appears and I will guess your age by adding the first number of the corresponding columns. Is this magic?

2	8	4	1
3	9	5	3
6	10	6	5
7	11	7	7
10	12	12	9
11	13	13	11
14	14	14	13
15	15	15	15

35. Product

The product of four consecutive whole numbers is 3024. What are these numbers?

36. Riverside Drive

Would you like to have dinner with me tonight? Don't go to the wrong house. I live in one of the 11 houses along Riverside Drive. When I am at home, facing the river, and I multiply the number of houses on my left by the number of houses on my right, I get a number which is greater by 5 units than the number that my neighbor to my left would get if he did the same thing. Where along the Drive do I live?

37. How old is Florence?

It's up to you to find out, knowing that the fifth power of her age less her age is a multiple of 10.

38. How old is the eldest?

You tell me, knowing that the square of the age of the middle boy is equal to the product of the ages of his brothers and that the sum of their three ages is 35, whereas the sum of the natural logarithms of their ages is 3.

39. How old am I?

Find the answer, knowing that Alfred's age can be found by inverting the order of the numerals that give my age; that when my age is divided by the sum of its two numerals, the result differs from the number obtained by performing the same operation on Alfred's age, by the difference of the two numerals; and that the product of these two quotients is exactly my age.

40. A question of years

On a nice Sunday in spring, a father goes on a walk with his sons. "Have you noticed," he asks them, "that the age of the eldest of you is equal to the sum of the ages of his two brothers?"

"Yes," they answer together, "and we have also noticed that if we multiply our ages together and by your age, we get a number that is equal to the sum of 1000 times the cube of the number of sons you have, added to 10 times its square."

Can you deduce from this conversation how old the father was when his middle son was born?

41. Ski lift

Three sisters are waiting in line for a ski lift, accompanied by their instructor. The teacher, getting bored, asks the eldest girl for their respective ages. "See if you can figure it out," she replies. "If you take away nine times the age of my youngest sister from the product of my age by the age of my other sister, you will get 89." The instructor, remembering that the middle sister had just received a pair of new skis for her tenth birthday, easily determined the ages of the other two girls. Can you do the same?

42. Summit meeting

Two delegations are to meet at the top floor of a skyscraper whose elevators can hold nine people at a time. The first delegation to arrive makes up a certain number of elevator loads, filling each one except the last, which has space for five more people. The second delegation does the same, not using one-third of the last elevator load.

At the start of the meeting, each member of each delegation shakes hands with each member of the other delegation, and each time, a photo is taken. Knowing that the photographer was using films with nine exposures on each, how many unexposed frames will he have left on his last film?

43. Rue Saint-Nicaise

On December 24, 1800, Napoleon Bonaparte, who was then First Consul, went to the opera using the Rue Saint-Nicaise. A bomb was set off in this street but a few seconds too late to harm him. There were a number of people killed and wounded. Bonaparte blamed the Republicans, 98 of whose numbers were deported to the Seychelles and Guyana. He also had some of them executed. Knowing that the number of wounded equals twice the number of killed (by the bomb explosion, not by the firing squads) added to four-thirds of the number of executions; that the total number of people killed, wounded, or executed is slightly less than the number of people deported; and that the number of dead less 4 is equal to twice the number of executions, can you determine, without reference to history books, how many Republicans were executed by Bonaparte as a result of the Rue Saint-Nicaise outrage?

44. A bag of marbles

A group of children share marbles from a bag. The first child takes one marble and a tenth of the remainder. The second child takes two marbles and a tenth of the remainder. The third child takes three marbles and a tenth of the remainder. And so on until the last child takes whatever is left. Knowing that all the children end up with the same number of marbles, how many children were there and how many marbles did each one get?

45. Saint Margaret's Academy

Boys are only admitted in the last grade of Saint Margaret's Academy. Their number is equal to the sum of the numerals representing the total number of pupils. On Saint Margaret's feast day, a celebration is due to take place in the chapel. Will it be possible to accommodate all the girls on benches designed to take nine girls at a time?

46. The emperor's soldiers

General Lasalle did not fear death: "Every Hussard more than 30 years old is a couldn't-care-less," he was fond of saying. Before one terrible battle, he counted his soldiers. The numerals representing their total added up to 17. After the battle, he added up the dead and the wounded. The numerals representing their number also added up to 17. The survivors marched away in full ranks of nine. How can you explain this?

47. Suez and Panama

A meeting takes place between Egyptians and Panamanians to discuss the operation of their respective canals. There are 12 of them in all, the Egyptians being more numerous than the Panamanians. The Egyptians arrive first, exchanging greetings, two by two, in their own language. The Panamanians arrive next and do the same. The Panamanians do not greet the Egyptians, or vice versa. Knowing that 31 greetings are exchanged in all, how many Egyptians were there at the meeting and how many Panamanians?

48. Number series

To write down all numbers 1 to n inclusive, it is necessary to use 2893 numerals. What is n?

49. Finding the trick

Maurice was the despair of all his teachers, particularly the mathematics teacher. However, when in order to solve a problem, he had to find the square of 35 or 75 or 85, he gave the right answer in 1 second flat. He had a trick of which he was very proud: To find the square of any two-digit number ending in 5, he multiplied the numeral representing the tens by this same number increased by 1 and wrote 25 to the right of the number thus obtained. For example, for 75^2, he would give his answer as "(7×8) 25" or 5625. Explain how Maurice's trick worked.

50. The virtues of the number 9

What is the last digit of $N = 99999^{9999^{999^{99^{9}}}}$?

51. Methuselah's greetings

Each year, from the year 1 of our era, Methuselah, who is still alive, sends greetings to his best friend. Although this best friend changes over the years, of course, the greetings remain the same: "Happy New Year 1," "Happy New Year 2," . . . , "Happy New Year 1982." Which is the least used numeral in this series?

52. The five numbers

Can you find five consecutive whole numbers that are all positive and such that the sum of the squares of the two largest numbers is equal to the sum of the squares of the three smallest?

53. Two sisters

Two sisters are 4 years apart in age. If one takes away the cube of the age of the younger sister from the cube of the age of the older, one gets 988. How old are they?

54. Fourth of July

A general notes one day that one can determine the number of soldiers in a garrison by adding to the number of barracks they occupy three times the square of this same number, then twice its cube. He then decides to hold a parade on July 4 with the soldiers marching in full ranks of six. What do you think of this decision?

55. A century old among them

Patrick goes walking with his father and his grandfather. They discuss the year when their three ages will add up to 100. The father and grandfather speak as follows:

The father: "I shall then be 28 years older than you will be and your age will be six times one-fifth of your present age."

The grandfather: "And I will be twice as old as your father would have been when you were born if you had been born a year and a half later than you actually were."

Patrick is puzzled by these statements and wonders how many years will elapse before they will have 100 years among them. Help him.

56. 100!

How many zeros are there are at the end of 100! ?

57. 1968 to 1978

Show that

$$N = \frac{7^{1968^{1978}} - 3^{68^{78}}}{1978 - 1968}$$

is a whole number.

Number secrets revealed

1. Barking

ROWFROWF = $(1 \times 4^0) + (3 \times 4^1) + (0 \times 4^2) + (2 \times 4^3)$
$+(1 \times 4^4) + (3 \times 4^5) + (0 \times 4^6) + (2 \times 4^7)$
$= 1 + 12 + 0 + 128 + 256 + 3072 + 0 + 32{,}768$
$= 36{,}237.$

2. The demonstration

Let x be the number of demonstrators.

(x + 1) must be a multiple of 2, 3, 4, . . . , 9.

(x + 1) is therefore a multiple of the smallest common multiple of these numbers, that is, a multiple of

$$2^3 \times 3^2 \times 5 \times 7 = 2520$$

Let $x + 1 = k(2520)$, k being a whole number.

$x = k(2520) - 1$. Since there are fewer than 5000 demonstrators, $k = 1$. Hence $x = 2519$, the number of marchers.

3. *Year of birth*

Let T, h, t, and u be the four digits that make up your year of birth (Thousands, hundreds, tens, and units).

Your year of birth can be written:

$$1000T + 100h + 10t + u.$$

If you take away

$$T + h + t + u$$

you are left with

$$999T + 99h + 9t$$

which is divisible by 9.

4. *Arithmetic*

Let us number the columns as follows:

```
    6 5 4 3 2 1
    T W E N T Y
  + T W E N T Y
  + T W E N T Y
  +       T E N
  +       T E N
  -------------
  = E I G H T Y
```

Let "col i" be the abbreviation for "column i."

Let "$r(x)$" be the carry of x, that is, the "tens" digit of the whole number x.

Let "$r(\text{col } i)$" be the "tens" digit of column i to be added to column $(i + 1)$.

Note I—Column 1:

$$2N + 3Y = Y + 10 \cdot r \text{ (col 1)}$$

Hence:

$$N + Y = 5, 10, \text{ or } 15$$

And:

$$r(\text{col } 1) = 1, 2, \text{ or } 3$$

Note II—Column 2:

$$3T + 2E + r(\text{col } 1) = T + 10 \cdot r(\text{col } 2)$$

Hence:

$$2T + 2E + r(\text{col } 1) = 10 \cdot r(\text{col } 2)$$

Hence:

$$r(\text{col } 1) \text{ is even}$$

Therefore:

$$r(\text{col } 1) = 2, N + Y = 10$$

and

$$T + E = 5 \cdot r(\text{col } 2) - 1$$

or

$$T + E = 4.9 \text{ or } 14$$

But (column 6) we have:

$$3T < 10; \text{ hence } T \leqslant 3$$

Hence:

$$T + E = 4 \text{ or } 9$$

Note III—Column 6:

$$E = 3T + r(3W)$$

But

$$T + E = 4 \text{ or } 9$$

hence

$$4T + r(3W) = 4 \text{ or } 9$$

But

$$r(3W) \leq 2;$$

hence the two remaining possibilities:

1. $4T + r(3W) = 9$

and therefore

$$r(3W) = 1 \text{ and } T = 2$$

$$E = 6 + 1 = 7; r(3E) = 2$$

and therefore (column 5):

$$3W + 2 = I + 10; 3W = I + 8$$

Hence:

$$I = 1/4 \text{ or } 7$$

But E =7; therefore, $I = 1$ or 4 and $W = 3$ or 4

If $W = 4$, $I = W$, which is impossible.

If $W = 3$, $I = 1$, $T = 2$, $E = 7$, and N or $Y = 4$ or 6.

If $N = 4$, $H = 8$, $Y = 6$, $G = 2$, which is impossible.

If $N = 6$, $H = 4$, $Y = 4$, which is impossible.

It is therefore impossible that $4T + r(3W) = 9$.

2. $4T + r(3W) = 4$

and therefore

$$r(3W) = 0 \text{ and } T = 1$$
$$E = 3; r(3E) = 0$$

and

$$r(\text{col } 4) = 0 \text{ or } 1$$
$$3W + (0 \text{ or } 1) = I + 10 \cdot r(\text{col } 5) = I + 10(0 \text{ or } 1).$$

We must therefore have $W \leq 3$. But 1 and 3 have already been assigned. And if $W = 0$, $I = 0$ also, which is impossible. The only possible value for W is therefore 2.

If $N = 6$, one can see (column 3) that $H = 1$, which is impossible.

If $N = 4$, $H = 5$, $G = 0$, and $Y = 6$, this is possible.

Check:

$$
\begin{array}{rrrrrrr}
 & 1 & 2 & 3 & 4 & 1 & 6 \\
+ & 1 & 2 & 3 & 4 & 1 & 6 \\
+ & 1 & 2 & 3 & 4 & 1 & 6 \\
+ & & & & 1 & 3 & 4 \\
+ & & & & 1 & 3 & 4 \\
\hline
= & 3 & 7 & 0 & 5 & 1 & 6
\end{array}
$$

5. Historical arithmetic

Let N be any year. $N = 1000T + 100h + 10t + u$, where

T is the "thousands" digit

h is the "hundreds" digit

t is the "tens" digit

u is the "units" digit

$$
\begin{aligned}
N &= (999 + 1)T + (99 + 1)h + (9 + 1)t + u \\
&= (999T + 99h + 9t) + (T + h + t + u)
\end{aligned}
$$

Dividing N by 9, we get

$$N/9 = (111T + 11h + t) + (T + h + t + u)/9$$

It can be seen that whatever the permutations of T, h, t, and u, N, when divided by 9, will always give the same remainder.

Therefore, any two years that possess the first characteristic given necessarily also possess the second. You cannot help the historian.

6. Spanish arithmetic

1. Let us examine the first column. $C = 1$ since $5C$ must be less than 10. Hence $V > 5$ (depending on the carry from the fifth column).

2. Examine the last column. $E = 5$ or 0. But E also appears in the second

column. Since $5U$ is a multiple of 5, there cannot have been a carry from the preceding column. Therefore, $A = 0$ (U cannot be 1 because $C = 1$). Therefore, $E = 5$ and both U and O must be odd.

3. Examine the third column. I must be the carry from $5T$.

But $I \neq (0 \text{ or } 1)$; hence $I = 2, 3,$ or 4 and $T > 3$.

If $U = 9$, then $V = 5 +$ carry from $(5 \times 9) = 9$.

U and V cannot both be 9; therefore $U \neq 9$.

That leaves 3 or 7 as the odd numbers possible for U.

Hence $V = 6$ or 8 and $O = 3, 7,$ or 9.

4. If $I = 4$, $T = 9$ (see 3 above); hence $O = 9$ (from an examination of columns 5 and 6), which is impossible.

5. If $I = 3$, $T = 6$ or 7. Hence in the fifth column, $T = 5 + (1 \text{ or } 2)$; hence $O = 3$. But $I = 3$, so this is impossible. Therefore, $I = 2$, $T = 4$, and $O = 9$.

Hence R is even; $R = 6$ or 8.

If $R = 8$, $N = 4 = T$, which is impossible. Therefore, $R = 6$.

Therefore, $V = 8$ and $U = 7$.

Proof: $170{,}469 \times 5 = 852{,}345$.

7. Graduation procession

Let a be the number of physical education teachers.

Let b be the number of Russian teachers.

Let N be the total number of pupils in the school.

We have

$$N = a \cdot b \cdot (a^2 - b^2) = a \cdot b \cdot (a + b) \cdot (a - b)$$

If a or b were divisible by 3, the principal could obviously organize a procession in full ranks of 3.

If neither a nor b were divisible by 3, there are four possibilities:

- a = multiple of $(3 + 1)$, b = multiple of $(3 + 2)$, hence $(a + b)$ is a multiple of 3.

- a = multiple of (3 + 1), b = multiple of (3 + 1), hence $(a - b)$ is a multiple of 3.
- a = multiple of (3 + 2), b = multiple of (3 + 1), hence $(a + b)$ is a multiple of 3.
- a = multiple of (3 + 2), b = multiple of (3 + 2), hence $(a - b)$ is a multiple of 3.

In all cases, therefore, N is a multiple of 3 and the pupils can parade in full ranks of three.

8. *Anniversary*

Let D be the initial number of dollars.

Let C be the initial amount of cents.

Let D' be the number of dollars that was left after the meal.

Let C' be the number of cents that was left after the meal.

Since Mr. Smith only had one-fifth of his money left,

$$100 \cdot D + C = 5(100 \cdot D' + C')$$

On the other hand, we were also told that $C' = D$ and $D' = C/5$.

From these three equations with four unknowns we derive that

$$C = (95/99)D$$

The number of cents, however, must be a whole number less than 100.

The initial number of dollars can therefore only be 99.

Hence $C = 95$, $D' = C/5 = 19$, $C' = D = 99$.

Mr. Smith therefore started out with $99.95 and ended with $19.99.

The meal cost him $79.96.

9. *The bus*

Let n be the number of passengers at the start of the journey.

Let n' be the number of passengers that got on at the first stop.

Let us work out, step by step, the number of passengers at any given instant, as a function of n and n'.

From the start to the first stop:

$$n$$

From the 1st stop to the 2nd:

$$(n) + (n') - (n') = n$$

From the 2nd to the 3rd:

$$(n) + (n + n') - (n') = 2n$$

From the 3rd to the 4th:

$$(2n) + (n + 2n') - (n + n') = 2n + n'$$

From the 4th to the 5th:

$$(2n + n') + (2n + 3n') - (n + 2n') = 3n + 2n'$$

From the 5th to the 6th:

$$(3n + 2n') + (3n + 5n') - (2n + 3n') = 4n + 4n'$$

From the 6th to the 7th:

$$(4n + 4n') + (5n + 8n') - (3n + 5n') = 6n + 7n'$$

From the 7th to the 8th:

$$(6n + 7n') + (8n + 13n') - (5n + 8n') = 9n + 12n'$$

From the 8th to the 9th:

$$(9n + 12n') + (13n + 21n') - (8n + 13n') = 14n + 20n'$$

From the 9th to the 10th:

$$(14n + 28n') + (21n + 34n') - (13n + 21n') = 22n + 33n'$$

Hence

$$22n + 33n' = 55$$

Hence $n = 1$ and $n' = 1$.

Between the seventh and eighth stops, therefore, there were 21 passengers on board ($9n + 12n' = 21$).

10. A memorable birthday

We know that Lucy's age is equal to four times the sum of the numerals that make it up. She must therefore be less than 100 years old (and more than 10).

Let t be the digit for the tens and u the digit for the units.

We have $10t + u = 4(t + u)$. Hence $2t = u$.

The digit representing the units is therefore twice the tens digit.

There are therefore only four possibilities: 12, 24, 36, and 48.

Since Lucy is over 40, she must have had her forty-eighth birthday on December 28, 1981. Thus her fortieth birthday was celebrated on December 28, 1973.

11. A story of 1s and 2s

$$\underbrace{11\ldots1}_{2n\text{ times}} - \underbrace{22\ldots2}_{n\text{ times}} = \underbrace{11\ldots1}_{n\ldots}\ \underbrace{11\ldots1}_{n} - 2\,[\underbrace{11\ldots1}_{n}]$$

$$= \underbrace{11\ldots1}_{n}\ \underbrace{00\ldots0}_{n} - \underbrace{11\ldots1}_{n}$$

$$= \underbrace{11\ldots1}_{n} \cdot \underbrace{100\ldots0}_{n} - 1$$

$$= \underbrace{11\ldots1}_{n} \cdot \underbrace{99\ldots9}_{n}$$

$$= \underbrace{11\ldots1}_{n} \cdot 9 \cdot \underbrace{11\ldots1}_{n}$$

$$= 3^2 \cdot \underbrace{11\ldots1}_{n}{}^{2}$$

$$= \underbrace{33\ldots3}_{n}{}^{2}$$

Example: $11 - 2 = 9 = 3^2$.

12. Great-grandfather's family

Let n be the unknown number. Great-grandfather tells us that

$$11 \times n \times 91 = 1001n = \text{“}nn\text{”}$$

This only tells us that n is a two-digit number between 10 and 99. Great-grandfather therefore has been extremely vague without appearing to be so.

13. Caramels

Let x be the number of caramels remaining to each child.

Let y be the number of caramels eaten by the oldest child (the third).

Number of caramels eaten by the second child: x.

Number of caramels eaten by the fourth child: $y + x + y$.

Hence the total number of caramels eaten is

$$y + x + y + (x + 2y) = 2x + 4y$$

On the other hand, we know that there are $4x$ caramels remaining.

Hence $4x + (2x + 4y) = 26$; thus $6x + 4y = 26$.

But x and y are whole, positive numbers. It is easy to conclude, therefore, that $x = 3$ and $y = 2$.

The eldest child therefore was given 5 caramels, the second 6, the third 5, and the fourth 10.

14. Square

We first note that the square of a whole number must end in a 0, 1, 4, 5, 6, or 9. In addition, the square of a whole number is a multiple of 4 (if the number is even), or a multiple of 4 with the addition of 1 if the number is odd.

We note then that the numbers ending in 11, 55, 66, or 99 are neither multiples of 4 nor multiples of 4 with 1 added.

This leaves us the numbers that end in 44, and therefore in 444.

A quick examination shows that whereas 444 is not a perfect square, 1444 is the square of 38.

There *are* therefore numbers whose square ends in three identical digits, other than 0.

15. Martin and Mildred

Let x be Martin's age, y that of Mildred.

What Martin says to Mildred can be summarized as

$$x = 3[y - (x - y)] \qquad \text{hence } 2x = 3y$$

The answer that Mildred gives to Martin can be written as

$$x + [x + (x - y)] = 77 \qquad \text{hence } 3x = y + 77$$

Substituting for y in the first equation gives us

$$2x = 3(3x - 77) \qquad \text{hence } x = 33$$

Hence $y = 3x - 77 = 22$.

Martin is 33 years old and Mildred is 22.

16. How many children were you?

The number of my sisters must be a multiple of 3 and a multiple of 2. It must therefore be a multiple of 6.

Total number of children: [(multiple of 6) × (1 + ½)] + 1.

Since this number must be less than 19, the only possible solution for the multiple of 6 is 6 exactly. I therefore have 6 sisters and 3 brothers. We are 10 children in all.

17. Airline company

Let n be the number of towns serviced by the airline.

Corresponding number of flights: $n(n - 1)$, since each town is joined to $(n - 1)$ other towns in the network.

Number of flights next year: $(n + k)\cdot(n + k - 1)$, k being the number of towns added to the network.

We therefore have

$$(n + k)\cdot(n + k - 1) - n\cdot(n - 1) = 76 = 4 \times 19$$

Hence

$$k(k + 2n - 1) = 4 \times 19$$

Thus either

$$k = 2 \text{ or } k = 4$$

If $k = 2$, $k + 2n - 1 = 38$; hence $2n - 1 = 36$, which is impossible because $(2n - 1)$ is an odd number.

Hence we must have $k = 4$

Thus

$$k + 2n - 1 = 19 \text{ and } n = 8$$

Eight towns are therefore serviced at present, and 12 will be serviced next year.

18. Latin exam

Let h be the highest of the two marks and s the smallest.

We have

$$h - (s/3) = 3(s - s/3) \qquad \text{hence } h = 7(s/3)$$

If $h = 7$ and $s = 3$, Michael has less than 50% of the highest possible mark (10). The only possible solution is therefore: $h = 14$ and $s = 6$. Hence Michael's mark was 14 and Claude's was 6.

19. Counting by fingers

The number of fingers on the left is $n - 1$.

The number of fingers on the right is $10 - n$.

But we have

$$9n = 10(n - 1) + (10 - n)$$

$n - 1$ must represent the digit for tens and $10 - n$ the digit for units.

20. Cluck-cluck

Let us start by computing the 10 first powers of 4:

1, 4, 16, 64, 256, 1024, 4096, 16,384, 65,536, 262,144.

The last power of 4 does not fit into 181,425. Hence $C = 0$.

4^8 goes twice into 181,425. Hence $L = 2$.

In the remainder (50,530), 4^7 goes three times. Hence $U = 3$.

4^6 does not go into the next remainder (1378); hence $C = 0$ again. But 4^5 goes into it once; hence $K = 1$.

Note: 181,425 in base 10 is written 230102301 in base 4.

21. Horse race

Left after the sixth race: 0.

Left after the fifth race: \$60.

Left after the fourth race: 60/2 = \$30.

Left after the third race: 30 + 60 = \$90.

Left after the second race: 90/2 = \$45.

Left after the first race: 45 + 60 = \$105.

Amount on hand before the first race: 105/2 = \$52.50.

I had \$52.50 when I arrived at Bowie racetrack.

22. Age difference

Let $Thtu$ be John's year of birth (Thousands, hundreds, tens, and units). Similarly, let $T'h't'u'$ be Jack's year of birth.

John's age in 1982: $1982 - (1000T + 100h + 10t + u)$.

Jack's age in 1982: $1982 - (1000T' + 100h' + 10t' + u')$.

Age difference: $1000(T - T') + 100(h + h') + 10\,(t - t') + (u + u')$.

But we know that $T + h + t + u = T' + h' + t' + u'$.

Hence $(T - T') + (h - h') + (t - t') + (u - u') = 0$.

Subtracting this from the age difference, we get

$$999(T - T') + 99(h - h') + 9\,(t - t')$$

which is obviously divisible by 9. Since this age difference must be less than 10 (the two ages start with the same digit), it must be 9. John and Jack differ in age by 9 years.

23. Division

Let D be the dividend, d the divisor, q the quotient, and r the remainder. The given information translates into

$$D = q \cdot d + r$$

$$D + 65 = q(d + 5) + r$$

Hence, by subtraction, $65 = 5q$ and hence $q = 13$.

24. Number switching

Let h, t, and u be the digits for the hundreds, the tens, and the units. We have

$$100h + 10t + u = (100h + 10u + t) - 45$$
$$100h + 10t + u = (100t + 10h + u) + 270$$

Hence

$$9t - 9u + 45 = 0$$
$$90h - 90t - 270 = 0$$

Hence

$$u = t + 5,\ h = t + 3$$

By interchanging the hundreds digit with the digit for the units, we get

$$(100u + 10t + h) - (100h + 10t + u) = 99(u - h)$$
$$= 99\,[(t + 5) - (t + 3)] = 198$$

The number therefore increases by 198.

25. School fire

Let c be the number of pupils in each classroom before the fire.

Let n be the number of classrooms before the fire.

We therefore have the following relationships:

$$nc = (n - 6)(c + 5)$$
$$nc = (n - 16)(c + 20)$$

Hence

$$-6c + 5n - 30 = 0$$
$$-16c + 20n - 320 = 0$$

Hence

$$c = 25 \text{ and } n = 36$$

The number of pupils in this school is therefore

$$nc = 36 \times 25 = 900$$

26. Brother and sister

Let x be the number of boys and y the number of girls.

What the boy says to the girl can be written as $x = y - 1$.

The answer that the sister gives her brother can be written as

$$y = 2(x - 1) \quad \text{or} \quad y = 2x - 2$$

Substituting $x = y - 1$ gives us $y = 4$ and hence $x = 3$.

There are therefore seven children in the family.

27. Large families

Let m and d be the numbers of children in the Martin and Davis families, respectively.

We have

$$m^2 - d^2 = 24; \text{ that is, } (m + d)(m - d) = 2^3 \times 3$$

Hence the various possibilities:

$m + d$	$m - d$	$2m$	m	d
24	1	25 : impossible	—	—
12	2	14	7	5
8	3	11 : impossible	—	—
6	4	10	5	1 : impossible

The only possible breakdown of $(m^2 - d^2)$ is therefore $m = 7$ and $d = 5$.

The Martins have seven children.

Note: The first two impossibilities result from the fact that $2m$ must be an even number. The third impossibility results from the assumption that there are no single children.

28. Eleanor's riddle

Let us assume that today is the day of my conversation with Eleanor.

Let Eleanor's age be e and my age be m.

Eleanor's age when I was her present age:

$$e - (m - e) = 2e - m$$

Her first sentence says that: $e = 3(2e - m)$ or $3m = 5e$
My age when Eleanor is three times as old as she is now:

$$m + (3e - e) = m + 2e$$

Eleanor's second sentence therefore says that

$$3e + (m + 2e) = 100 \qquad \text{or} \qquad 5e + m = 100$$

Combining this with her first pronouncement gives us

$$3m + m = 100 \qquad \text{or} \qquad m = 25$$

Hence

$$e = 3m/5 = 15$$

Eleanor was therefore 15 years old.

29. Ursula and the cats

Let n be the number of cats.

We have $n = 4/5 \cdot n + 4/5$. Hence $n = 4$.

Ursula lives with four cats.

30. New math

Let a be the unknown base. When my son writes 253 in base a, this can be interpreted as

$$2a^2 + 5a + 3 = 136$$

or

$$2a^2 + 5a - 133 = 0$$

or

$$(a - 7)(2a + 19) = 0$$

Since a is a whole, positive number, the only possible solution is 7; the unknown base is 7.

31. Multiplication

```
     5283
 ×     49
    47547
   21132
   258867
```

32. Chinese numbers

Any number n equal to or smaller than 26 must certainly be smaller than $3^3 = 27$. It can therefore be broken down (in base 3) in the following manner:

$$n = (0 \text{ or } 1 \text{ or } 2)\cdot 3^0 + (0 \text{ or } 1 \text{ or } 2)\cdot 3^1 + (0 \text{ or } 1 \text{ or } 2)\cdot 3^2$$

In other words

$$n = (0 \text{ or } 1 \text{ or } 2) + (0 \text{ or } 3 \text{ or } 6) + (0 \text{ or } 9 \text{ or } 18)$$

Thus all numbers that have an "1" in their makeup will be found in the first square, all numbers that have a "2" in the second square, a "3" in the third square, a "6" in the fourth square, a "9" in the fifth, and an "18" in the sixth.

The number n can therefore be found by adding the numbers at the top left-hand corner of the squares where n appears.

33. *The smallest possible*

Let n be this unknown number. Since n divided by 2 leaves a remainder of 1, $n + 1$ must be divisible by 2.

Since n divided by 3 leaves a remainder of 2, $n + 1$ must be divisible by 3. Similarly, $n + 1$ must be divisible by 4, 5, and 6.

The smallest common multiple of 1, 2, 3, 4, 5, and 6 is 60.

Therefore, $n + 1 = 60$. Hence $n = 59$.

34. *For those under 16*

It is not a matter of magic but simply the characteristics of a number written in base 2. Any whole number less than 16 can be written as

$$x_0 \cdot 2^0 + x_1 \cdot 2^1 + x_2 \cdot 2^2 + x_3 \cdot 2^3$$

where each x_n is either a 0 or a 1.

When $x_n = 1$, the corresponding age is in the column headed 2^n.

When $x_n = 0$, the corresponding age is *not* in the column headed 2^n.

Example: If you are 13 years old, your age will appear in the last three columns because

$$13 = 1 \cdot 2^3 + 1 \cdot 2^2 + 0 \cdot 2^1 + 1 \cdot 2^0 = 8 + 4 + 0 + 1$$

35. *Product*

3024 ends in neither a 5 nor a 0; therefore, none of the four numbers given can be divisible by 5 or by 10. If all four numbers were greater than 10, their product would exceed 10,000, which is not the case. The four numbers must either be 1-2-3-4 or 6-7-8-9. In the first case, the product is 24. The answer can only be 6-7-8-9, as can be easily verified.

36. *Riverside Drive*

Let r be the number of houses to the right of my house and l the number of houses to the left of my house when I look toward the river. We have the following simultaneous equations:

$$r + l = 10$$

$$rl - (r + 1) \cdot (l - 1) = 5$$

Hence $r + l = 10$ and $r - l = 4$;

thus $r = 7$ and $l = 3$.

My house is therefore the fourth from the left when facing the river.

37. How old is Florence

Let a be the unknown age. We have $a^5 - a = 10k$ (k being any whole number).

Hence

$$a(a^2 - 1)(a^2 + 1) = 10k \qquad \text{or} \qquad a(a + 1)(a - 1)[(a - 2)(a + 2) + 5] = 10k$$

Whatever a turns out to be, either a or $a + 1$ is even and one of the five numbers a, $a + 1$, $a + 2$, $a - 1$, or $a - 2$ is a multiple of 5. Therefore, a^5a is a multiple of 10 for all values of a. Therefore, we have no clue as to Florence's age, which can be any number between 0 and 110 years!

38. How old is the eldest?

Let the eldest boy be a years old, the middle one c years old, and the youngest one b years old. We have

$$c^2 = ab$$

$$a + b + c = 35$$

$$\log a + \log b + \log c = 3$$

Taking logarithms of the first equation yields

$$2 \log c = \log a + \log b$$

Hence, from the third equation, $\log c = 1$.

Thus $c = 10$; the middle boy is 10 years old.

Hence $ab = 100$ and $a + b = 25$; a and b are the roots of $x^2 - 25x + 100 = 0$. Since $a > b$, $a = 20$ and $b = 5$. The eldest boy is 20 years old.

39. How old am I?

Let x be my age, t being the digit representing the tens and u the digit representing the units:

$$x = 10t + u\text{; Alfred's age: } 10u + t$$

The first proposition says that

$$\left|\frac{10t+u}{t+u} - \frac{10u+t}{t+u}\right| = |t-u|$$

or

$$9\,|t-u| = (t+u)\cdot|t-u| \qquad \text{or} \qquad t = 9-u$$

The second proposition says that

$$\frac{10t+u}{t+u} \cdot \frac{10u+t}{t+u} = 10t+u$$

or

$$10u + t = (t+u)^2$$

Substituting $t = 9 - u$, we have $9u + 9 = 81$.

Hence $u = 8$ and $t = 1$.

I am 18 years old. Alfred is 81 years old.

40. A question of years

There are three sons. The product of their ages is

$$1000\cdot3^3 + 10\cdot3^2 = 27{,}090 = 43 \times 7 \times 5 \times 3^2 \times 2$$

The eldest being as old as the ages of his brothers added together, the only possible breakdown is 43-14-9-5.

The father is 43 years old; the middle son is 9 years old.

The father was therefore 34 when his middle son was born.

41. Ski lift

The eldest must be more than 10 years old, the middle sister is exactly 10, and the youngest sister must be less than 10.

Let the eldest be a years old and the youngest b.

We have

$$10a - 9b = 89 \qquad \text{or} \qquad 10(a - b) + b = 10(8) + 9$$

Hence

$$a - b = 8 \text{ and } b = 9$$

The ages of the three sisters are 17, 10, and 9.

42. Summit meeting

Let d_1 be the number of the members in the first delegation and d_2 the number of members in the second.

$$d_1 = (\text{multiple of } 9) + 4$$
$$d_2 = (\text{multiple of } 9) + 6$$

Number of photographs:

$$\begin{aligned} d_1 \times d_2 &= (\text{multiple of } 9) + (4 \times 6) \\ &= (\text{multiple of } 9) + (18 + 6) \\ &= (\text{multiple of } 9) + 6 \end{aligned}$$

The photographer therefore exposed six frames on his last film.

There are three frames left.

43. Rue Saint-Nicaise

Let x be the number of executions.

Number of killed in the bomb explosion: $2x + 4$.

Number of wounded: $2(2x + 4) + 4/3x$ or $5x + x/3 + 8$.

x is therefore a multiple of 3.

We also have the following relationship:

$$(2x + 4) + (5x + x/3 + 8) + (x) < 98 \qquad \text{or} \qquad x < 10$$

x must therefore be equal to 9: Bonaparte had nine Republicans executed after the bomb attempt of the Rue Saint-Nicaise.

44. A bag of marbles

Let n be the number of children, x each child's share of marbles, and N the total number of marbles. We have $N = nx$.

First child's share: $1 + (N - 1)/10 = x$.

Last child's share: x.

Penultimate child's share: $n - 1 + x/9 = (N/x) - 1 + x/9 = x$.

Considering the expression for the first child's share: $N = 10x - 9$.

Substituting into the equation for the penultimate child's share gives us

$$(10x - 9)/x - 1 + x/9 = x$$

which can be written

$$8x^2 - 81x + 81 = 0 \qquad \text{or} \qquad (x - 9)(8x - 9) = 0$$

The number of marbles taken by each child being a whole number, the only possible solution is $x = 9$;

hence $N = 81$ and $n = 9$.

There were nine children. Each got nine marbles.

45. Saint Margaret's Academy

Let us suppose that the total number of pupils attending this school is less than 10,000, which seems reasonable. Thus the number can be written as *Thtu*, where T is the digit representing the thousands, h the hundreds, t the tens, and u the units.

Total number of pupils: $1000T + 100h + 10t + u$.

Number of boys: $T + h + t + u$.

Number of girls: $999T + 99h + 9t + 0$ (difference between the two previous numbers). This last number is therefore a multiple of 9.

The girls will all fill out benches built to seat nine.

46. The emperor's soldiers

Let us suppose first, as seems fair enough, that General Lasalle had fewer than 10,000 soldiers.

Let $Thtu$ be this number. $N = 1000T + 100h + 10t + u$.

We have: $T + h + t + u = 17$.

Number of dead and wounded: $T'h't'u'$.

$$N' = 1000T' + 100h' + 10t' + u' \text{ with } T' + h' + t' + u' = 17$$

Hence

$$(T - T') + (h - h') + (t - t') + (u - u') = 0 \qquad (1)$$

Let the number of uninjured survivors after the battle be

$$U = 1000(T - t') + 100(h - h') + 10(t - t') + (u - u') \qquad (2)$$

Combining (1) and (2):

$$U = 999(T - T') + 99(h - h') + 9(t - t')$$

This is divisible by 9. The survivors could therefore march away in full ranks of nine.

47. Suez and Panama

When n people meet, the number of greetings exchanged is $n(n - 1)/2$, hence the following table:

Number of Panamanians	Number of Egyptians	Number of Greetings		Total Greetings
		Panamanians	Egyptian	
1	11	0	55	55
2	10	1	45	46
3	9	3	36	39
4	8	6	28	34
5	7	10	21	31

In order to have a total of 31 greetings, it was necessary to have seven Egyptians and five Panamanians at the meeting.

48. Number series

In order to write the first 9 numbers with one digit, it is necessary to use $9 \times 1 = 9$ numerals.

In order to write the next 90 numbers that have two digits, it is necessary to use 180 numerals.

In order to write the next 900 numbers that have three digits, it is necessary to use 2700 numerals.

Hence a total of 2889 numerals are required to write all numbers up to 999.

In order to write all numbers up to n, we require 2893 numerals = 2889 + 4. Therefore n must be the first four-digit number: that is, 1000.

49. Finding the trick

Let t be the digit representing the tens of the number to be squared. This number can also be written $10t + 5$.

If this is squared, we will get $100t^2 + 100t + 25$, which can also be written $100t(t + 1) + 25$.

For example:

$$85^2 = 100(8 \times 9) + 25 = 7225$$

This is the explanation for Maurice's trick.

50. The virtues of the number 9

We note first that

$$999^{99999^9}$$

is an odd number. We can therefore express it as $(2k + 1)$, where k is a whole number.

N can be written

$$99999^{2k+1} \text{ or } 11111^{2k+1} \times 9^{2k+1} \text{ or } 11111^{2k+1} \times 81^k \times 9$$

But 11111^{2k+1} ends in a 1. So does 81^k. Therefore, their product does, too. Therefore, nine times their product ends in a 9.

N must also end in a 9.

51. Methuselah's greetings

From 1 to 999, all numerals have been used the same number of times except for 0 (111 fewer times). Similarly, from 1000 to 1999, all the numerals have been used the same number of times except for 1 (1000 times more often). But we are only considering the set up to 1982. There remains 1983, 1984, 1985, . . ., 1999. In this new set, it is the numeral 9 that appears most often (29 times), but this lack of 9s in the main set is of a lesser magnitude than the lack of 0s noted in the first 1000 numbers.

0 is therefore the numeral used least often.

52. The five numbers

Let n be the number in the middle of the series. We have the relationship

$$(n + 1)^2 + (n + 2)^2 = n^2 + (n - 1)^2 + (n - 2)^2$$

Expanding and simplifying gives us

$$n(n - 12) = 0$$

The only acceptable solution is $n = 12$. The five consecutive whole numbers are

10-11-12-13-14

53. Two sisters

Let the elder be x years old.

Let the younger be y years old.

We have

$$x = y + 4$$
$$x^3 - y^3 = 998$$

Hence $12y^2 + 48y + 64 = 988$.

Thus $y = 7$ and $x = 11$.

54. Fourth of July

Let s be the number of soldiers and b the number of barracks.

We have

$$s = b + 3b^2 + 2b^3 = b(b + 1)(2b + 1)$$

It is obvious that $b(b + 1)$ is divisible by 2, whatever b is.

On the other hand, if b is a multiple of 3, so is s.

If b is a (multiple of 3) + 1, $(2b + 1)$ is a multiple of 3.

If b is a (multiple of 3) + 2, $(b + 1)$ is a multiple of 3.

s is therefore a multiple of 3 and a multiple of 2 whatever b is.

It is therefore a multiple of 6; the general's decision can be implemented.

55. A century old among them

Let c be the age of the child.

Let f be the age of the father.

Let g be the age of the grandfather.

Let x be the time that has to elapse before their ages add up to 100. We have the following relationships:

$$(g + x) + (p + x) + (c + x) = 100$$

$$(c + x) = (6/5)c$$

$$(p + x) - (c + x) = 28$$

$$g + x = 2(p - c + 1{\cdot}5)$$

From the last two equations, we deduce that $g + x = 59$.

Hence, from the first equation:

$$59 + (p + x) + (c + x) = 100$$

Hence $p + x = 28 + (c \times x$ and $c\ x = (6/5)c$.

From this we conclude that

$$59 + 28 + (12/5)c = 100$$

Hence $c = 5$ years and 5 months and $x = c/5 = 1$ year and 1 month.

Their ages will add up to 100 in 13 months' time.

56. 100!

The number of zeros at the end of a number is equal to the number of times that number is a multiple of 10. However, 10 is a factor of 5 and 2. The number of zeros will therefore be equal to the smaller of the following two numbers: number of times 2 appears as a factor and number of times 5 appears as a factor in the breakdown of the original number into prime factors.

Here, of course, 2 appears as a factor more often than 5. Let us calculate the number of times 5 appears as a factor: 100! has 20 numbers that are multiples of 5. Some of them are even multiples of 25: 25, 50, 75, and 100. In the breakdown of 100! into prime factors, one will find 5 raised to the power 24 $(20 + 4 = 24)$. There are therefore 24 zeros at the end of 100! .

57. 1968 to 1978

Let n be any whole number. The last digit of 7^n is 7, 9, 3, or 1. If n is a multiple of 4, the last digit is 1. This is because $7^{4k} = (7^4)^k = (2401)^k$.

The last digit of 3^n is 3, 9, 7, or 1. If n is a multiple of 4, the last digit is 1. This is because $3^{4k} = (3^4)^k = (81)^k$.

It is obvious, on the other hand, that 1968 and 68 are multiples of 4. Each part of the numerator therefore ends with the numeral 1. Their difference must therefore end in a zero. The numerator is therefore a multiple of 10 and N is a whole number.

6. Exercises in Cunning and Common Sense

1. British hospital

In a British hospital, the following information was posted: "Surgeon *X* and nurse *Y* are pleased to announce their forthcoming marriage." Knowing that, when the hospital opened, the fiancé was as old as the fianceé is now and that the product of the three ages (that of the two betrothed and of the hospital) less that of surgeon *X* is 1539, when was the hospital opened?

2. The hands of my watch

It is later than 3:20 but not yet 3:25. I note the exact position of the hands on my watch. I then turn the hands, without harming the watch, in order to position the large hand where the small hand used to be and the small hand where the large one used to be. What is the exact time?

3. At the restaurant

The first course of a three-course meal is served between 7 and 8 o'clock, when the two hands of the restaurant clock are equidistant from the number 6. The dessert arrives when the big hand has caught up with the small hand. How long did it take to eat the first and second courses?

4. Livestock epidemic

A farmer owned an equal number of cows, pigs, horses, and rabbits. After an epidemic, the following cries of despair were uttered by the various members of the family:

The farmer: "One cow in every five died."

The farmer's wife: "As many horses as pigs survived."

The farmer's son: "The ratio of rabbits to total number of animals is now 5/14."

The farmer's mother: "Death has not spared a single species of animal."

But the farmer's mother was very old. Can you demonstrate that her statement is incorrect?

5. Zoo talk

If in all the zoos that house hippos and rhinos there are no giraffes; if in all the zoos that house rhinos and that have no giraffes there are hippos; and if in all the zoos where there are hippos and giraffes there are rhinos, can there be a zoo where there are hippos but no rhinos or giraffes?

6. Coffee break

We go out for coffee between 1 and 2 o'clock when the bisector of the angle formed by the hands of my watch passes through noon. What is the exact time when we leave the office?

7. Bats, bears, and Chinamen

A careful observation of physical phenomena has shown that 17 bears eat as much as 170 Chinamen, 100,000 bats eat as much as 50 Chinamen, and 10 bears eat as much as 4 elephants. How many bats will eat as much as a dozen elephants?

8. House of cards

Do you know how to build card houses?

The first floor is easy:

The second floor is easy, too:

Here's the third:

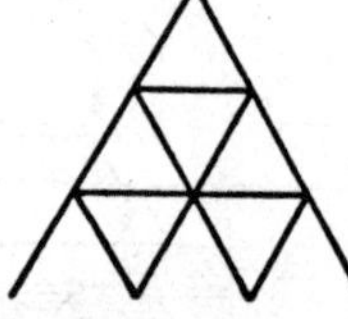

And so forth.

How many cards does it take to build a 47-story house?

9. Cigarettes

Four couples dine together. During the evening, Diana smokes three cigarettes, Elizabeth smokes two, Nicola smokes four, and Maude smokes one. Simon smokes as many as his wife, Peter twice as many as his wife, Lewis three times as many as his wife, and Christopher four times as many as his spouse. Knowing that 32 cigarettes are smoked in all, what is the first name of Lewis's wife?

10. Confession

A confessor was in the habit of giving specific penances for specific sins. Thus, before giving absolution for the sin of pride, he would ask the sinner to recite one "Hail Mary" and two "Lord's Prayer." In the case of libel, the penance would be two "Lord's Prayer" and seven "Credo." For the sin of sloth, he would require 2 "Hail Mary," for adultery 10 "Hail Mary," 10 "Lord's Prayer," and 10 "Credo." For the sin of gluttony, the penance would be one "Hail Mary." Atoning for the sin of selfishness would require three "Lord's Prayer" and one "Credo." The penance for covetousness was three "Lord's Prayer." Finally, atonement for slander necessitated seven "Lord's Prayer" and two "Credo." Knowing that at my last confession I admitted to having committed 12 sins for which I had to recite 9 "Hail Mary," 12 "Lord's Prayer," and 10 "Credo," what was the nature of my sins?

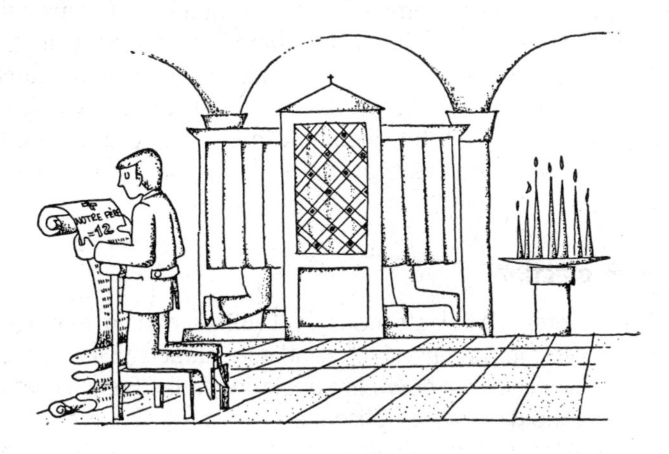

11. Typists

Two typists are asked to type a text. The first completes the first half and the second does the rest. They complete the job in 25 hours. Knowing that had they worked simultaneously they would have completed the job in 12 hours, how long would each have taken had she done the job alone?

12. Lacemakers

Two lacemakers must complete a piece of lace. If they work alone, the first will take 8 days to finish the work and the second will take 13. How long will they require to do the job together?

13. Lunch with friends

Ten couples meet for lunch. After a cocktail they go into the dining room in a random order, one by one. How many people must enter the dining room to assure that:

1. There is at least one married couple in the group?
2. There are at least two diners of the same sex?

14. Slumbering camels

Some camels sleep, some camels don't. The number of camels that sleep is equal to seven-eighths of the number of camels that don't sleep plus seven-eighths of a camel. If half of the sleeping camels stopped sleeping, the number of nonsleeping camels would be between 25 and 65. If all the sleeping camels ceased to do so, the number of camels that don't sleep would be . . . Please complete this sentence.

15. French election

In the French presidential elections of 1988, a certain number of candidates are in the running. Each candidate attracts twice as many votes as the next candidate down in popularity. In the French election system, if no candidate has an absolute majority, a runoff must take place between the two candidates with the most votes. Is a runoff necessary here?

16. Grandmother, Latin test, and kittens

If you write down twice, side by side, the mark I got on my Latin test (scored from 0 to 10), you will obtain the age of my grandmother. If you divided this age by the number of kittens I own, you will obtain three times my Latin grade plus fourteen-thirds. Can you tell me the age of my grandmother?

17. Mathematical puzzle

One week, the puzzle in "Today's Mathematician" was particularly difficult. When I started to work on it, between 9 and 10 o'clock, the two hands of my watch opposite one another in a straight line. I was unable to solve the puzzle until the hands of my watch coincided. How long did the puzzle-solving process take?

18. Seals

I recently went to the aquarium to see the seals. There weren't many seals. In fact, there were only seven-eighths of the total number of seals plus seven-eighths seals. How many seals could I see?

19. Subway

On a certain subway line, trains leave regularly every 10 minutes from each end of the line throughout the day and take exactly 1 hour to reach their destination at the other end. A commuter gets on at one of the end stops and travels the whole distance. How many trains will pass him going in the opposite direction?

20. Highwaters

A Parisian parking lot on the banks of the Seine contained a number of Renaults, Peugeots, and Citroëns. There were twice as many Renaults as Peugeots, the latter being twice as numerous as the Citroëns. Following torrential rains, the Seine overflowed its banks and carried away many cars. There were as many Citroëns out of the water as there were Renaults in the water, three times as many Renaults out of the water as there were Peugeots in the water, and as many Peugeots out of the water as Citroëns out of the water. How many Citroëns did the Seine carry away?

21. Toy figures

Nicholas owns an equal number of toy Indians, Arabs, cowboys, and Eskimos. After a visit by his cousin Sebastian, he is furious to discover that one-third of his figures have disappeared. Knowing that there are as many Eskimos remaining as there are cowboys missing, and that he still has two-thirds of his Indians, how many Arabs did Sebastian steal?

22. Office cleaning

The offices of a large corporation are distributed over three floors of a tall building: the 13th and 14th floors, where the general offices are located, and the 25th floor, where management resides. A team of cleaners is called in to do the spring cleaning. For 4 hours they tackle the general offices. Then they split into two groups, the first half moving up to the 25th floor while the others remain where they are. At the end of another 4 hours, the team stops work. The 13th and 14th floors are finished, and only 8 hours of work by a single cleaner is needed to complete the 25th floor. Given that the time required for cleaning is proportional to the floor area, how many people are there on the cleaning team?

23. Inhabitants

I live in a city with 2,754,842 inhabitants. If each person were assigned a number between 1 and 2,754,842, how many numerals would be required to write down all these numbers? What would the sum of these numbers be, and what would be the sum of all the numerals used to write down these numbers?

24. Loose change

I had $10 in my purse in coins of 50, 10, and 5 cents. But now I have nothing because I gave my dimes to Victor, my half-dollars to Margaret, and my nickels to George. Who do you think got the least money, knowing that one of the two boys got nine times as many coins as the other?

25. Grade crossings

Two small villages, *A* and *B*, are on the same straight portion of a railroad track, 10 km apart. In spite of the flat country, the road that joins them zigzags. A cyclist observes that no matter what his position *P* on this road, the following relationship is true (the projection of *P* on the railroad track being *H*): "The sum of 20 times the length *BH*, of 13 times the cube of this same length, and of 300 times the length *PH* is equal to the sum of the length *BH* raised to the fourth power and of 32 times the square of this same length (all lengths being expressed in kilometers)." How many grade crossings will the cyclist encounter between *A* and *B*?

26. Tribute

Every year, a king expects each of his 30 vassals to give him 30 gold pieces. He knows, however, that one vassal has gotten into the unfortunate habit of giving him 9-gram coins instead of the 10-gram coins required. How can the king, in a single weighing operation, identify the culprit in order to chop off his head?

27. Drugstores

A new town has 13 blocks, *a, b, c, d, . . ., l, m*. We know that *a* is contiguous to *b* and *d*; *b* to *a, c,* and *d*; *c* to *b*; *d* to *a, b, f,* and *e*; *e* to *d, f, j,* and *l*; *f* to *d, e, j, i,* and *g*; *g* to *f, i,* and *h*; *h* to *g* and *i*; *i* to *j, f, g,* and *h*; *j* to *k, e, f,* and *i*; *k* to *l* and *j*; *l* to *k* and *e*; whereas *m* is an isolated block. Each inhabitant of this new town must be able to find a drugstore nearby, either in his or her own block on in a neighboring block. Where will the drugstores have to be located?

28. Fill 'em up

A foreigner has two cars—one large, one small. Their gas tanks have a combined capacity of 70 liters. When both tanks are empty, it costs her 45 crowns to fill one tank and 68 crowns to fill the other. Knowing that the small car uses regular gas and the large car uses premium gas and that where the foreigner lives there is a 20-cent (1 cent = 1 hundredth of a crown) difference between the price of regular and premium, what are the capacities of her two cars' gas tanks?

29. Apples

Audrey picks up fallen apples, eats one-fourth of them, and distributes the rest among her small sisters, Bertha, Charlotte, and Dorothy. Bertha eats one-fourth of her apples then gives the rest in equal shares to the other three girls. Charlotte eats an apple, then divides the remainder equally among her sisters. Dorothy eats four apples before following her sister's example. "Look," Bertha says to Charlotte, "I have twice as many apples as you do." How many apples did Audrey pick up?

30. Buying power

While prices went up 12%, Mr. X's salary went up 22%. By how much did his buying power increase?

31. Childish psychology

A small boy spends most of his time painting. His paintings always represent either an Indian or an Eskimo, in all cases alongside some sort of dwelling, usually igloos for the Eskimos and wigwams for the Indians. But sometimes the opposite is true and, for example, Indians will be standing by igloos. A psychologist studies this case and finds that there are twice as many Indians as Eskimos, as many Eskimos with wigwams as Indians with igloos, and that for every Eskimo alongside a wigwam, three are standing by igloos. The psychologist then tries to find the proportion of Indians among the wigwam dwellers (in order to determine, by a very complex formula, the nature of the boy's Id). But she is unable to do so. Can you help her?

32. Time, please?

In order to find out, add the time that remains until noon to two-fifths of the time that has elapsed since midnight. . . .

33. The reservoir

The reservoir has the shape of a rectangular parallelepiped whose width equals half its length. It is filled to three-eighths of its height. By adding 76 hectoliters, the level climbs by another 0.38 meter.

Two-sevenths of the reservoir remains to be filled. What is the reservoir's height?

34. Family reunion

Nine members of the same family meet. Each member arrives alone but at the same time. For very complex psychological reasons that we will not explore, each member kisses five others in the family and shakes the hand of three more. Where is the contradiction?

35. Parent/teacher meeting

At the end of the school year, the tenth-grade teachers meet with a certain number of class parents. At one of these meetings it was noted that a total of 31 people were present. The Latin teacher was questioned by 16 parents, the French teacher by 17 parents, the English teacher by 18 parents, and so on up to the mathematics teacher, who was questioned by all the parents present. How many parents attended the meeting?

36. Savages

A certain tribe of savages reckons that 1 year is made up of 12 months and that 1 month is made up of 30 days. Another tribe reckons that 1 year is made up of 13 moons, each moon of 4 weeks, and each week of 7 days. The two tribes meet in a democratic fashion to elect a common chief. The first tribe wants to empower the new leader for 7 years, 1 month, and 18 days; whereas the second tribe wants to elect him for a term of 6 years, 12 moons, 1 week, and 3 days. Which of these two periods is longer?

37. Flour sacks

Two trucks were carrying identical sacks of flour from France into Spain. The first truck was delivering 118 sacks, the second only 40. Lacking enough Spanish currency to pay the import duty, the first truck driver decided to give the customs officer 10 sacks of flour. This left him with 800 pesetas to pay. The second driver followed suit, handing over 4 sacks, for which he *received* 800 pesetas from the customs officials. Knowing that the exact amount of import duty was paid to the customs officials, how much was each sack of flour worth?

38. Sheila's neuroses

Sheila was extremely neurotic. She went to see a therapist, who succeeded after lengthy treatment in curing her of half of her neuroses plus half of one more. She then went to a second therapist, who succeeded in curing her of half her remaining neuroses plus an additional half. A third therapist did the same. Sheila was left with only one neurosis (which remained for the rest of her life). Knowing that each therapist charged $197 per cured neurosis, what was the total cost of the treatment?

39. A square plot of land

Matthew owns a square plot of land. A large track of constant width with an area of 464 m^2 forms the inside perimeter. When he makes a circuit of his field, Matthew notices that there is a difference of 32 meters between the circuits, depending on whether he follows the inside edge or the outside edge of the track. What is the total area of the plot of land?

40. Tennis tournament

One hundred ninety-nine players sign up for a tennis tournament. Lots are drawn to determine who shall play whom for the first round, the 199th player joining in on the second round only. On this second round, lots are also drawn for the matches. The same is true for round three, and so on (each time there is an odd number of players, one player skips that round). Knowing that a can of new balls is required for each match, how many cans will be needed for the entire tournament?

41. Thermometer

A defective thermometer reads +1°C in melting ice and +105°C in the steam of boiling water. What is the real temperature when the thermometer reads +17°C?

42. The barrel

In how many ways can a 10-liter barrel be emptied by means of two containers of 1-liter and 2-liter capacities, respectively?

43. Old Ihsan

Old Ihsan dwells in a small Turkish village whose inhabitants frequently live to a ripe old age. Ihsan has an extremely large family, 2801 in all, including his children, grandchildren, great-grandchildren, great-great-grandchildren, and himself. Each family member has had the same number of children (except the great-great-grandchildren, who have not yet procreated) and none of them have died. Can you tell me how many children Ihsan has?

44. Calves, cows, pigs, and chickens

The dowry of a young peasant girl consists of a cow, a calf, a pig, and a chicken. Her fiancé remembers that five cows, seven calves, nine pigs, and a chicken cost a total of 108,210 francs. He also knows that a cow costs 4000 francs more than a calf, that three calves cost as much as 10 pigs, and that 3000 chickens cost as much as five calves. Given this information, he tries to calculate the value of his beloved's dowry, but fails. Can you help him?

45. 32 cards

Alfred, Brian, Christopher, and Damon play with a deck of 32 cards. Damon deals them out unequally, then says: "If you want us to have the same number of cards, do exactly as I say. You, Alfred, divide half of your cards between Brian and Christopher. Then, Brian, you do the same with Christopher and Alfred. Finally, Christopher, you follow suit with Alfred and Brian." How did Damon initially distribute the cards?

Elementary, my friend

1. British hospital

Let x = surgeon's age (fiancé or fiancée).

Let y = nurse's age (male or female).

Let h = hospital's age.

The information given can be translated into the following two equations:

$$|x - y| = h$$

$$xyh - x = 1539 \qquad \text{or} \qquad x(yh - 1) = 1539 = 3^4 . 19$$

Bearing in mind that x and y are the ages of adult people, the last equation can be satisfied only for the following whole numbers for x and $(yh - 1)$:

x	$yh - 1$
19	81
27	57
57	27
81	19

hence

yh	y	h
82	41	2
58	29	2
28	28	1
20	20	1

If we substitute these values of x, y, and h in the first equation, it will be found that this equation is satisfied only for the second solution of the four given in the table: that is, $x = 27$, $y = 29$, and $h = 2$.

The wedding is therefore between a 27-year-old woman surgeon and a 29-year-old male nurse in a hospital that opened 2 years ago.

2. The hands of my watch

If l and s are the angles formed by the hands of my watch, respectively, with noon direction, then

$$l = 12\left(s - h\frac{2\pi}{12}\right)$$

where h is the number that represents time.

If l_0 and s_0 correspond to the initial positions of the hands,

$$l_0 = 12\left(s_0 - \frac{\pi}{2}\right) \tag{1}$$

If l_1 and s_1 correspond to the new positions,

$$l_1 = 12\left(s_1 - \frac{2\pi}{3}\right)$$

but we know that $l_1 = s_0$ and $s_1 = l_0$; hence

$$s_0 = 12\left(l_0 - \frac{2\pi}{3}\right) \tag{2}$$

Solving (1) and (2) for l_0 gives us

$$l_0 = 12\left[12\left(l_0 - \frac{2\pi}{3}\right) - \frac{\pi}{2}\right]$$

Hence

$$l_0 = \frac{6 \times 17\pi}{143} \quad \text{or} \quad \frac{3 \times 17 \times 60}{143} \text{ minutes}$$

since 2π on the clock = 60 minutes.

The time at the start of the exercise is therefore

3 hours 21 minutes 24 seconds

3. At the restaurant

Let 7 hours x minutes be the time when the first course was served.

The angle between the small hand and the 6 o'clock line is

$$\frac{360}{12} + \left(\frac{x}{60} \times \frac{360}{12}\right) = 30\left(1 + \frac{x}{60}\right)$$

The angle between the large hand and the 6 o'clock line is

$$180 - x\,\frac{360}{60} = 180 - 6x$$

Since these two angles are equal,

$$x = \frac{300}{13} = 23 \text{ minutes } 5 \text{ seconds}$$

The first course was therefore served at 7 hours 23 minutes 5 seconds.

Let 7 hours y minutes be the time when the dessert arrived.

The angle between the small hand and the 6 o'clock line is

$$30\left(1 + \frac{y}{60}\right)$$

The angle between the large hand and the 6 o'clock line is

$$6y - 180$$

Since these two angles are equal,

$$y = \frac{420}{11} = 38 \text{ minutes } 11 \text{ seconds}$$

The dessert was therefore served at 7 hours 38 minutes 11 seconds.

It therefore took 15 minutes and 6 seconds to eat the first and second courses.

4. Livestock epidemic

Let N be the total number of animals before the epidemic.

From the information given by the farmer, the number of cows that survived is

$$(4/5)(N/4) = N/5$$

From the information given by the farmer's wife, the number of surviving pigs or horses is $N/4$.

Let n be the number of dead rabbits.

The information given by the son translates into

$$\frac{N/4 - n}{N/5 + N/4 + (N/4 - n)} = \frac{5}{14}$$

That is,

$$5(N - 4n)/(14N - 20n) = 5/14$$

which can be satisfied only for $n = 0$.

No rabbits died. The farmer's mother was wrong.

	Cows	Pigs	Horses	Rabbits
deaths	$N/20$	$N/4 - x$	x	n
	$N/5$	x	$N/4 - x$	$N/4 - n$
	$N/4$	$N/4$	$N/4$	$N/4$

N

5. *Zoo talk*

Yes. Explanation: Let Z be the set of zoos, H the subset of those with hippos, R the subset of those with rhinos, and G the subset of those with giraffes.

The first statement tells us that I is empty.

The second statement tells us that II is empty.

The third statement tells us that III is empty.

But in no way does this imply that IV (with hippos, but without giraffes or rhinos) is empty.

Q.E.D.

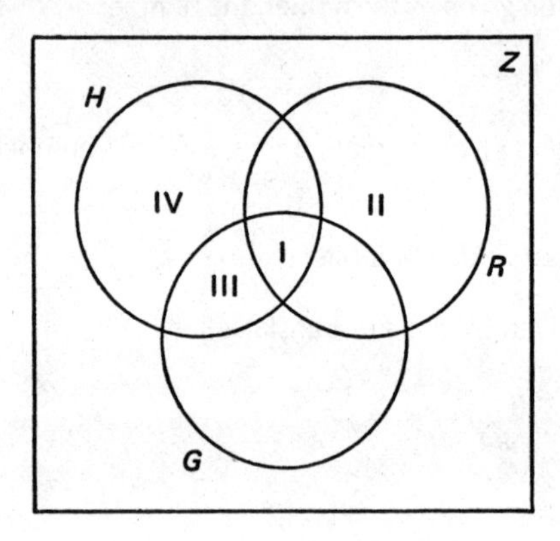

6. *Coffee break*

Let 1 hour x minutes be the unknown time. The angle of the small hand with the bisector (in degrees) is

$$(360/12) + (x/60)(360/12) = 30 + x/2$$

The angle of the large hand with the bisector is

$$360 - x(360/60) = 360 - 6x$$

Hence

$$30 + x/2 = 360 - 6x$$

That is,

$$x = (2/13)330 = 50 \text{ minutes } 46 \text{ seconds}$$

We leave the office at 1 hour 50 minutes 46 seconds.

7. *Bats, bears, and Chinamen*

Let y be the quantity absorbed by an elephant in the course of a day.

Let x be the quantity absorbed by a bat in the course of a day.

Let z be the quantity absorbed by a Chinaman in the course of a day.

Let t be the quantity absorbed by a bear in the course of a day.

$$17t = 170z \qquad 100{,}000\,x = 50z \qquad 10t = 4y$$

Hence

$$t = 10z \qquad z = 2000x \qquad y = 5/2t$$

Hence

$$y = (5/2)10 \cdot 2000x = 50{,}000x$$

It will take 600,000 bats to eat as much as a dozen elephants.

8. *House of cards*

Note first that in all houses of cards, every level has three cards more than the level immediately above it.

The number of cards for a 47-story house is therefore

$$\begin{array}{l} \quad 2 \\ + 2 + 3 \\ + 2 + 2.3 \\ + \ldots\ldots \\ + 2 + (n - 1) \,.\, 3 \\ + \ldots\ldots\ldots\ldots \\ + 2 + (47 - 1) \,.\, 3 \\ \hline 47 \cdot 2 + \dfrac{46 \cdot 47}{2} \cdot 3 = 3337 \end{array}$$

3337 playing cards are needed to build a card house 47 stories high!

Note: It may not be clearly apparent that

$$1 + 2 + 3 + \cdots + 46 = \frac{46 \cdot 47}{2}$$

The proof is that you can take the members of this arithmetic progression in pairs:

$$1 + 2 + 3 + \cdots + 23 + 24 + \cdots + 45 + 46$$

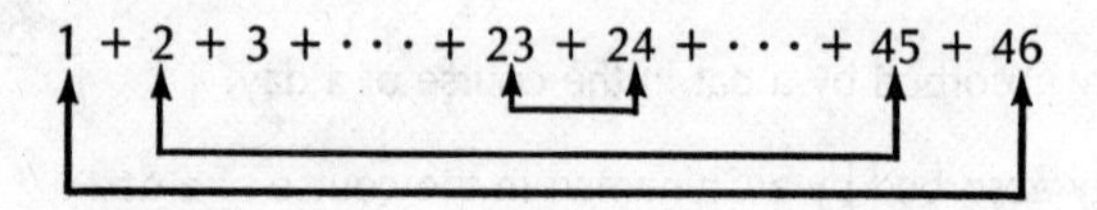

Each pair adds up to 47 and there are 46/2 pairs, hence

$$\frac{46 \cdot 47}{2} \ldots$$

9. Cigarettes

If 32 cigarettes were smoked in all, the four men took 22 (32 − 3 − 2 − 4 − 1).

Twenty-two can be broken up into (3 × 1) + (4 × 2) + (3 × 1) + (2 × 4) and no other way. Thus Lewis, who smoked three times as many cigarettes as his wife, must be married to Maude.

10. Confession

First: It should be noted that since I had to recite 9 "Hail Mary," I could not have been accused of adultery.

Second: I must have confessed to libel; otherwise, given that I had to recite 10 "Credo," I would have had to commit sins of selfishness and slander several times, which would have resulted in more than 12 "Lord's Prayer."

Third: After my penance for libel, I still had to recite three "Credo." This could be due to three sins of selfishness or a single sin of selfishness associated with one of slander. But three sins of selfishness would result in 9 "Lord's Prayer"; with the libel this would total 11. Since I had to recite 12 "Lord's Prayer," this leaves one unexplained and inexplicable. Therefore, my additional three "Credo" came from one sin of slander and one of selfishness.

Fourth: There remains to determine the nine sins corresponding to nine "Hail

Mary": they must therefore be nine sins of gluttony. I therefore confessed to one sin of slander, one of libel, one of gluttony, and nine of greed.

(It is recommended that the reader make a table summarizing the penances for each of the eight sins considered.)

11. Typists

Let x be the number of hours that the first typist would have taken to complete the work alone and y the corresponding number of hours for the second typist. To complete half the task, it would have taken the first typist $x/2$ hours and the second $y/2$ hours.

Hence

$$(x + y)/2 = 25$$

On the other hand, when they are working simultaneously, they carry out in one hour a portion of the work equal to

$$1/x + 1/y = (x + y)/xy$$

Hence

$$xy/(x + y) = 12$$

Hence

$$xy = 12(x + y) = 600$$

Since we now know the sum and the product of x and y, we know that they are the roots of the equation $x^2 - 50x + 600 = 0$.

Hence

$$x = 20 \text{ and } y = 30$$

It would have taken the first typist 20 hours and the second typist 30 hours to complete the work alone.

12. Lacemakers

In one day the first lacemaker completes one-eighth of the work and the second one-thirteenth.

Together the task will therefore take them

$$1 \div (1/8 + 1/13) = 4.95 \text{ days}$$

13. Lunch with friends

As soon as 11 people enter the dining room, there will necessarily be one married couple among them.

As soon as three people enter the room, there will necessarily be two diners of the same sex.

14. Slumbering camels

Let *s* be the number of camels that sleep and a the number of those that remain awake.

We have

$$s = 7/8\ a + 7/8 \qquad \text{or} \qquad 8s = 7a + 7$$

s must therefore be a multiple of 7.

In addition, since *s* can be divided into two equal parts that are whole numbers, *s* must be a multiple of 14.

Hence the following possible solutions:

s	14	28	42	56	etc.
a	15	31	47	63	etc.
a + s/2	22	45	68	91	etc.

If half the camels that sleep stop sleeping, the number of camels that are awake will be given by the third line in the table. If this number is to be between 25 and 65, it must be 45. Hence $s = 28$ and $a = 31$. The total number of camels is 59. This is the number we are looking for.

15. French election

Let *n* be the number of candidates.

Let x be the number of votes received by the preferred candidate.

The number of votes received by all the other candidates was

$$x\left(\frac{1}{2} + \frac{1}{4} + \ldots + \frac{1}{2^{n-1}}\right) = \frac{x}{2}\left(\frac{1 - 1/2^{n-1}}{1 = 1/2}\right) = x - \frac{x}{2^{n-1}}$$

The preferred candidate therefore received more votes than all the others combined. There was no need for a runoff.

16. Grandmother, Latin test, and kittens

Let a be the age of my grandmother.

Let c be the number of kittens.

Let l be my marks in the Latin exam.

According to the first sentence, l must be a whole number less than 10 and such that $a = 11l$.

The second sentence can be expressed

$$a/c = 3l + 14/3 \qquad \text{hence } a = 3cl + 14c/3$$

c is therefore a multiple of 3 such that $3c < 11$.

c must therefore be equal to 3.

We then have

$$a = 11l$$
$$a = 9l + 14$$

Hence

$$l = 7 \text{ and } a = 77$$

My grandmother is 77 years old.

17. Mathematical puzzle

In 12 hours, the large hand has turned 11 complete rotations more than the small one (or 22 half rotations more). Each half rotation will take 12 hours/22 = 32 minutes 44 seconds.

Hence the puzzle took me 32 minutes 44 seconds to solve.

18. Seals

Let n be the number of seals. We have

$$n = 7(n/8) + 7/8$$

Hence $n = 7$

There were seven seals in the aquarium.

19. Subway

The commuter will cross all the trains that left the other end of the line in the hour preceding his departure (generally six). He will also cross all the trains that will leave in the hour that follows his departure (six also, in general).

He will therefore cross 12 trains.

Special case: If the departures from both ends are simultaneous, our commuter will cross only 11 trains but he will see 2 more, one at each end of the line.

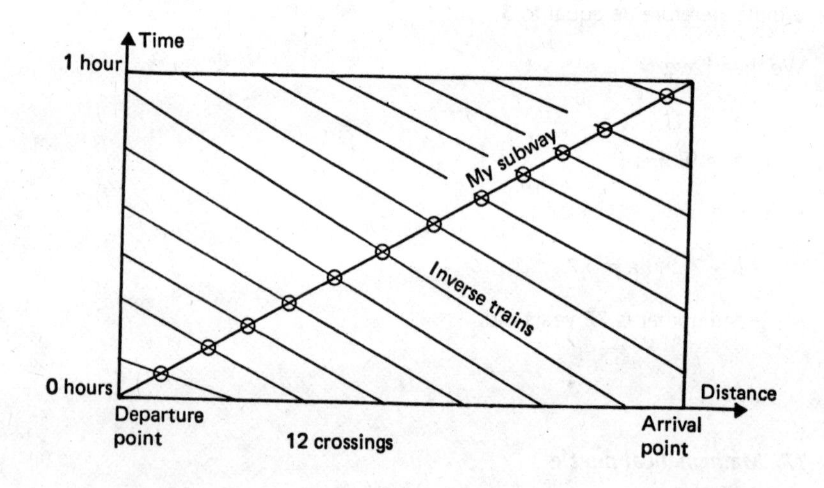

20. High waters

Let x be the total number of Citroëns.

Let y be the total number of Renaults carried away by the Seine.

Total number of Peugeots = $2x$.

Total number of Renaults = $2 \times 2x = 4x$.

Number of Renaults out of the water = $4x - y$.

Number of Peugeots in the water = $4x - y$.

Number of Citroëns out of the water = y.

Number of Peugeots out of the water = y.

Total number of Peugeots: $(1/3)(4x - y) + y = 2x$.

Hence: $x = y$.

Therefore, all the Citroëns stayed out of the water. The Seine did not carry off a single one of them.

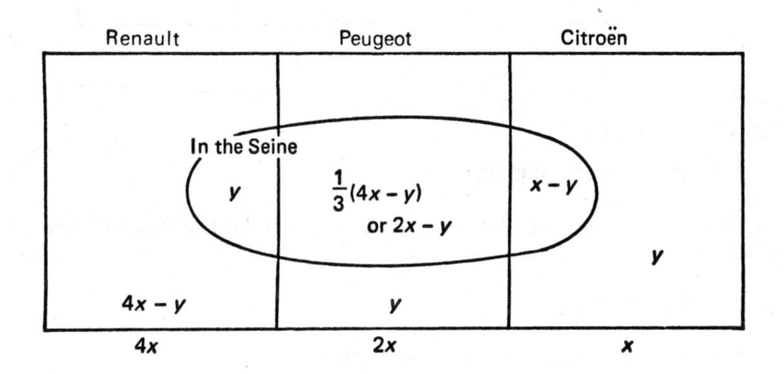

21. Toy figures

Let x be the number of figures of each type that Nicholas originally had in his collection.

Let y be the number of cowboys missing or of the remaining Eskimos.

Number of Eskimos missing: $x - y$.

We also know that the number of Indians missing is $x/3$.

Let z be the number of missing Arabs. The total number of missing figures is $4x/3$, as it is a third of the total number of figures.

But it is also $y + (x - y) + x/3 + z$.

Hence $4x/3 = 4x/3 + z$.

Thus $z = 0$.

Cousin Sebastian did not take any Arabs.

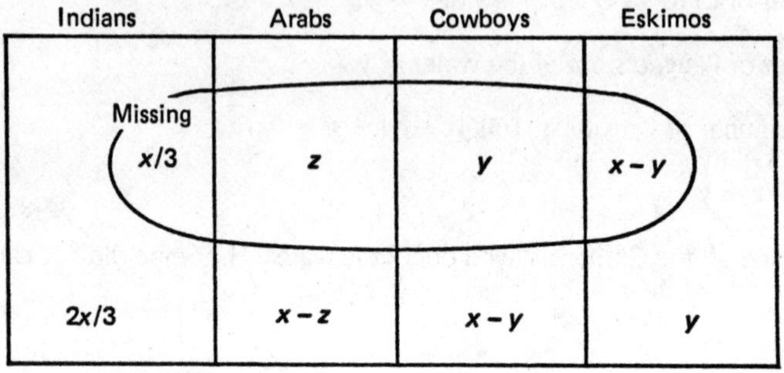

	Indians	Arabs	Cowboys	Eskimos
Missing	$x/3$	z	y	$x - y$
	$2x/3$	$x - z$	$x - y$	y

22. *Office cleaning*

Let n be this unknown number.

Number of hours required by one cleaner to clean the two floors of general offices:

$$4 \times (n + n/2)$$

Number of hours required to clean the 25th floor:

$$4 \times n/2 + 8$$

But this last number must be half of the preceding one because the corresponding area is half the other one.

Hence $4(n + n/2) = 2(4n/2 + 8)$.

Thus $n = 8$.

23. *Inhabitants*

1. Number of numerals: N_1

There are

9 numbers with 1 numeral = 9
90 numbers with 2 numerals = 180
900 numbers with 3 numerals = 2700
9000 numbers with 4 numerals = 36,000
90,000 numbers with 5 numerals = 450,000
900,000 numbers with 6 numerals = 5,400,000
1,754,843 numbers with 7 numerals = 12,283,901

Hence $N_1 = 18{,}172{,}790$.

2. Sum of the numbers: N_2

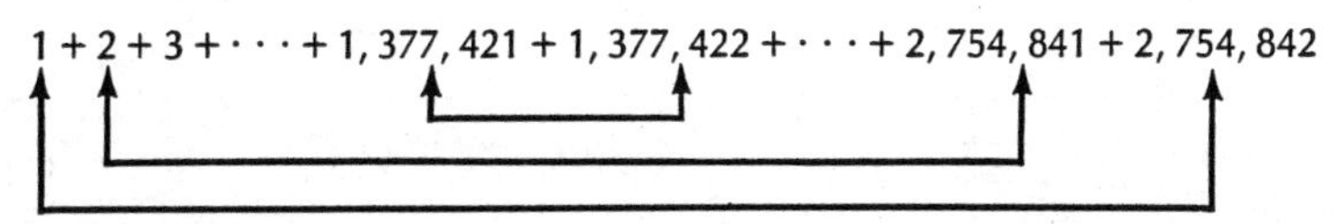

The pairs of numbers add up to 2,754,843. Hence

$$N_2 = 1{,}377{,}421 \times 2\ 2{,}754{,}843 = 3{,}794{,}579 \text{ million}$$

3. Sum of all the numerals: N_3.

To find the answer to this tricky question, we will present a condensed form of the solution.

Let $(n \rightarrow N)$ be the sum of the numerals of all the numbers between n and N inclusive.

A. Establish by recurrence that

$$(1 \rightarrow 10^n - 1) = n \cdot 45 \cdot 10^{n-1}$$

B. Deduce by dissection and summation that

$$(1 \rightarrow p \cdot 10^n - 1) = \frac{p}{2} \cdot 10^n(9n + p - 1)$$

Then

$$(1 \rightarrow p \cdot 10^n) = \frac{p}{2} \cdot 10^n(9n + p - 1) + p.$$

C. Apply all this to our example.

Slice	Direct application of the above formula	Sum of recurring numerals
1 to 2,000,000	55,000,002	
2,000,001 to 2,700,000	17,850,007	2 × 700,000
2,700,001 to 2,750,000	1,000,005	(2 + 7) × 50,000
2,750,001 to 2,754,000	60,004	(2 + 7 + 5) × 4000
2,754,001 to 2,754,800	10,008	(2 + 7 + 5 + 4) × 800
2,754,801 to 2,754, 840	244	(2 + 7 + 5 + 4 + 8) × 40
2,754,841 to 2,754,842	3	(2 + 7 + 5 + 4 + 8 + 4) × 2
Total: N_3 = 75, 841, 733		

This solution comes from readers of the French publication *Valeurs Actuelles.*

24. Loose change

Let x be the number of dimes.

Let y be the number of half-dollars.

Number of nickels:

either $x/9$ (case 1, Victor got more coins than George)
or $9x$ (case 2, George got more coins than Victor)

Case 1. The total value of the coins being \$10, we have

$$x + (x/9 \cdot 1/2) + 5y = 100$$

Hence

$$19x + 90y = 1800$$
$$19x = 1800 - 90y$$

Since both x and y are whole positive numbers, it can be seen that this is impossible.

Case 2. The total value of the coins is now

$$x + 9x/2 + 5y = 100$$

Hence

$$11x + 10y = 200$$

This time there is one possible solution: $x = 10$, $y = 9$. Hence Margaret and George each got \$4.50. Victor got only \$1.00. He lost out on the deal.

25. Grade crossings

Let x be the length BH and y the length PH.

The given relationship can be expressed as follows:

$$20x + 13x^3 + 300y = x^4 + 32x^2$$

or

$$\begin{aligned} 300y &= x^4 - 13x^3 + 32x^2 - 20x \\ &= x(x - 10)(x^2 - 3x + 2) \\ &= x(x - 10)(x - 1)(x - 2) \end{aligned}$$

PH therefore becomes zero at B and A, 1 km from B and 2 km from B.

The road from A to B therefore crosses the railroad track in two points, and the two corresponding grade crossings are 1 km and 2 km from B.

26. Tribute

It is sufficient to weigh a pile of gold pieces consisting of one piece from the first vassal, two pieces from the second, three from the third, and so on. If all the vassels give 10-gram pieces, the total pile will weigh

$$10(1 + 2 + 3 \cdots + 30) = 10 \frac{30(30 + 1)}{2} = 4650 \text{ grams}$$

If the actual weight is 1 gram short, the first vassal will be the guilty one, if 2 grams short the second will be guilty, and so on.

27. Drugstores

The drugstores must be located in blocks *b*, *i*, *l*, and *m*. This becomes obvious from the following sketch, where each block is represented by a point and the points are joined together if the blocks are contiguous.

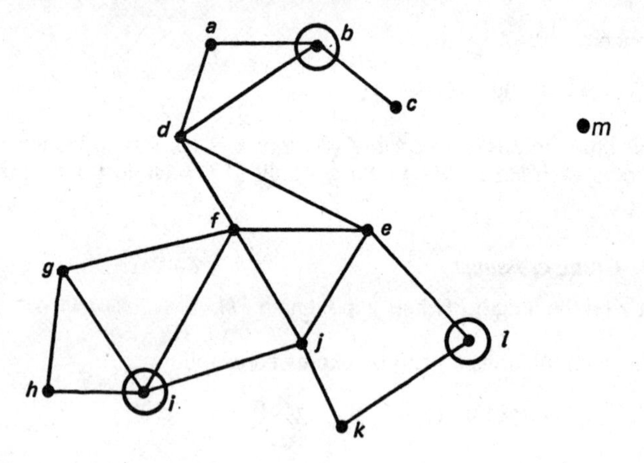

28. Fill 'em up

Let x be the tank capacity of the small car.

Let X be the tank capacity of the large car.

Let p be the price in crowns of 1 liter of regular gas.

We have

$$x + X = 70$$

$$xp = 45$$

$$X(p + 0.2) = 68$$

Eliminating p, we have

$$x + X = 70$$

$$45X + 0.2xX = 68x$$

Eliminating x gives us

$$2x^2 + 990x - 31{,}500 = 0$$

which can be written

$$(x - 30)(2x + 1050) = 0$$

The only possible solution is $x = 30$ liters
Hence $X = 40$ liters
These are the required tank capacities.

29. Apples

Let a be Audrey's final share, c Charlotte's final share, and $2c$ that of Bertha. (Dorothy has no apples left). By means of the following table, it is possible to work backward to arrive at the initial shares:

	Audrey	Bertha	Charlotte	Dorothy
End	a	$2c$	c	0
Before action by Dorothy	$a - c$	c	0	$4 + 3c$
Before action by Charlotte	$a - 2c$	0	$1 + 3c$	$4 + 2c$
Before action by Bertha or after action by Audrey	0	$4a - 8c$	$1 + 5c - a$	$4 + 4c - a$

These last shares must be equal. Hence we have the equations

$$4a - 8c = 1 + 5c - a$$

$$4a - 8c = 4 + 4c - a$$

Hence $c = 3$, $a = 8$.

Audrey therefore distributed 8 apples to each of her sisters. She had picked up 32 apples.

30. Buying power

Consider Mr. X's original salary. Without a raise, his buying power would have been reduced by the factor (1 + 0.12). With his raise, his buying power was multiplied by

$$(1 + 0.22)/(1 + 0.12) = 1.22/1.12 = 1.089$$

His buying power therefore increased by 8.9%.

31. Childish psychology

Let EW be the proportion of paintings representing Eskimos and wigwams.

Let EI be the proportion of paintings representing Eskimos and igloos.

Let II be the proportion of paintings representing Indians and igloos.

Let IW be the proportion of paintings representing Indians and wigwams.

We have the following relationships:

$$EW + EI + II + IW = 1$$

$$(IW + II) = 2\,(EW + EI)$$

$$II = EW$$

$$EI = 3EW$$

Eliminating II and EI yields

$$EW + 3EW + IW + EW = 1$$

$$(IW + EW) = 2(EW + 3EW)$$

Hence

$$5\,EW + IW = 1$$

$$7\,EW - IW = 0$$

and

$$EW = 1/12$$
$$IW = 7/12$$

The proportion sought by the psychologist is therefore

$$IW/(IW + EW) = (7/12) \div (8/12) = 7/8$$

32. Time, please?

Let h be the unknown time. It is the time that has elapsed since midnight since it is the present time. The time remaining to noon therefore is $12 - h$. The given information translates as

$$h = (12 - h) + 2h/5$$

Hence $h = 60/8$ hours.

It is therefore 7:30 a.m.

33. The reservoir

0.38 corresponds to the height left after having filled 3/8 of the total and before having still to fill 2/7 of the total. Hence, it is equal to $(1 - 3/8 + 2/7) =$ 19/56 of the total height. The required answer is therefore

$$(56/19 \cdot 0.38 = 1.19 \text{ meters}$$

Hence, the required answer is

$$56/19 \times 0.38^m = 1.12 \text{ meters}$$

34. Family reunion

The total number of handshakes given is equal to half the sum of the number of hands shaken by each member:

$$(1/2)(9 \times 3) = 27/2$$

One should find a whole number. Since the answer is a fraction, the data given are flawed.

35. Parent/teacher meeting

Let n be the number of parents present at the meeting. Let m be the number of teachers. The first teacher spoke to $15 + 1$ parents; the second spoke to $15 + 2$ parents, . . . ; the mth teacher spoke to $15 + m$. But the mth teacher was the math teacher, who spoke to all the parents.

Hence

$$15 + m = n$$

But 31 people were present; hence

$$n + m = 31$$

Subtracting the two equations, we have

$$n - 15 = 31 - n$$

Hence

$$n = 23$$

Twenty-three parents attended the meeting.

36. Savages

To compare the given time periods, they must be converted to days.

First time period:

$$(7 \times 12 \times 30) + 30 + 10 = 2568$$

Second time period:

$$(6 \times 13 \times 4 \times 7) + (12 \times 4 \times 7) + 7 + 3 = 2530$$

The first period is therefore the longest.

37. Flour sacks

Let x be the price of a sack of flour.

Let y be the import duty on each sack.

If the first trucker gave 10 sacks to the customs officer, he had to pay duty on only 108 sacks. Similarly, the second trucker had to pay duty on only 36 sacks.

From the data given,

$$10x + 800 = 108y$$
$$4x - 800 = 36y$$

Eliminating y, we have

$$2x - 3200 = 0$$

Hence

$$x = 1600$$

Each sack of flour was worth 1600 pesetas.

38. Sheila's neuroses

Let us summarize the data by means of a table:

Treatment period	Number of neuroses
At the end	1
Before the third therapist	$(1 + 0.5)2 = 3$
Before the second therapist	$(3 + 0.5)2 = 7$
Before the first therapist	$(7 + 0.5)2 = 15$

Sheila therefore started out with 15 neuroses. The total cost of the treatment was

$$14 \times 197 = \$2758$$

39. A square plot of land

Let x be the length of the outside of the track.

Let y be the length of the inside of the track.

We have the following equations:

$$x^2 - y^2 = 464/(x + y)(x - y) = 464$$
$$4x - 4y = 32/x - y = 8$$

Hence $x + y = 58$.

Thus $x = 33$ and $y = 25$.

The total area of Matthew's plot of land is 1089 m^2.

40. Tennis tournament

Each match involves the elimination of a player. Since there are 199 players, 198 will be eliminated. There will thus be 198 matches, and 198 cans of balls will be needed for the tournament.

41. Thermometer

Real temperature:

$$\frac{17 - 1}{(105 - 1)/100} = 15.38°$$

42. The barrel

Let u_n be the number of ways of emptying an n-liter barrel with two containers of 1-liter and 2-liter capacity, respectively.

Either one starts with the 1-liter container (and there will then be $n - 1$ liters remaining) or with the 2-liter container (there will then be $n - 2$ liters remaining).

Hence the recurring formula

$$u_n = u_n - 1 + u_n - 2$$

Here we obviously have $u_0 = 1$, $u_1 = 1$.

Hence

$$u_2 = 2$$
$$u_3 = 3$$
$$u_4 = 5$$
$$u_5 = 8$$
$$u_6 = 13$$
$$u_7 = 21$$
$$u_8 = 34$$

$$u_9 = 55$$
$$u_{10} = 89$$

There are thus 89 ways to empty a 10-liter barrel using the two containers.

43. Old Ihsan

Let the answer be n.

Number of grandchildren: n^2.

Number of great-grandchildren: n^3.

Number of great-great-grandchildren: n^4.

Total number in family group of himself and of direct descendants:

$$1 + n + n^2 + n^3 + n^4 = 2801$$

Hence

$$n + n^2 + n^3 + n^4 = 2800 = 2^4 \cdot 5^2 \cdot 7$$

Hence n, which is a factor of 2800, must be equal to either 2, 4, 5, 7, or 8.

Since $8^4 = 4096 > 2800$, $n < 8$.

A trial with seven shows that this is the correct solution.

Old Ihsan has seven children.

44. Calves, cows, pigs, and chickens

Let p be the price of a pig.

Price of a calf: $(10/3)p$.

Price of a cow: $(10/3)p + 4000$.

Price of a chicken: $(5/3000)(10/3)p$.

Hence

$$(50/3)p + 20{,}000 + (70/3)p + 9p + p/180 = 108{,}210$$

Thus $p = 1800$.

The dowry is therefore worth

$$6000 + 10{,}000 + 1800 + 10 = 17{,}810 \text{ francs}$$

45. 32 cards

At the end of the game each of the four players has 8 cards. Christopher, having just shared half his cards between Brian and Alfred, must have had 16 cards and Alfred and Brian 4 each. But Brian had just shared half his cards between Christopher and Alfred: before that share-out, Brian must have had 8, Alfred 2, and Christopher 14. But this was just after Alfred shared half his cards between Brian and Christopher.

Hence at the start Alfred had 4 cards, Brian 7, Christopher 13, and Damon (who was not involved in the various share-outs) 8.

7 And now, your turn

1. Leap frog

Consider seven squares in a line. Place three white counters and three black counters in the squares in the first drawing below. You may then move the counters in accordance with the following rules:

- Black counters can only be moved to the right
- White counters can only be moved to the left
- At each move, a counter can only move to an empty square, either by moving one square or by leapfrogging over a counter of the other color.

At the start:

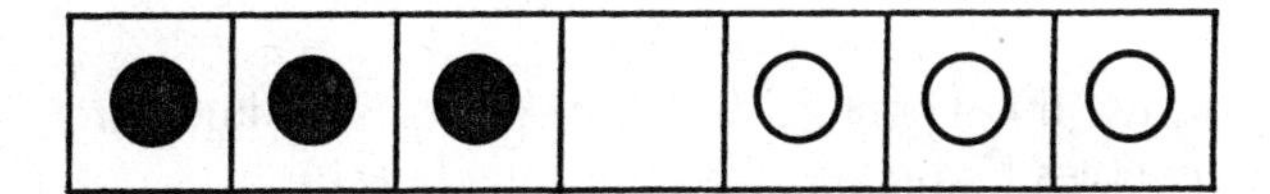

Example of a lost game: if you move black counters twice in succession at the start of the game, the only possible third move is to move the last black counter and no further moves are possible. The game is lost.

Lost game:

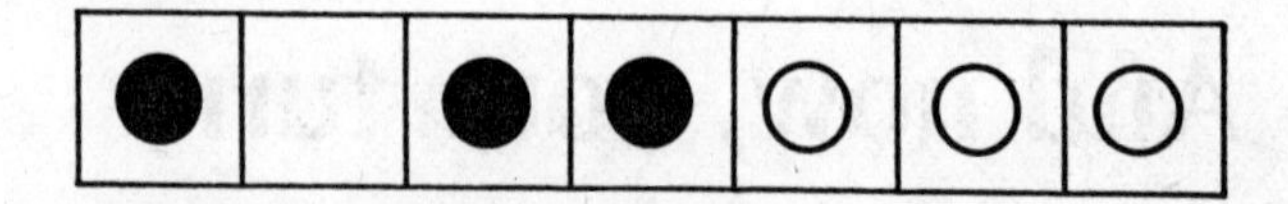

You will have won if you succeed in interchanging the white and the black counters. Now, your turn to play.

2. Game of the three roads

Mention of this ancient game, well-known in the Far East, appears in the work of Ovid where it is written: "This game is played on a special small table with three coins for each player. In order to win, it is necessary to get one's three coins in a straight line."

To play the game yourself, pencil the following figure onto a piece of cardboard or trace it into the sand with a stick:

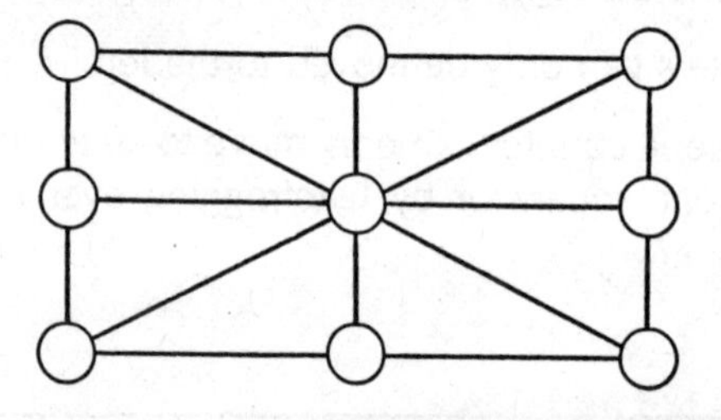

For counters use three pennies and three dimes (substitute three black and white pebbles if you are playing on the beach).

Choose a partner as intelligent as yourself and deal three similar coins to each of you. The player who starts the game places his or her first coin in one of the circles. The second player does the same, placing a coin in one of the remaining eight circles. The first player then places a second coin; and so on. When all the coins have been placed in position, each player, in turn, moves one of his or her three coins over one of the 16 segments, assuming of course that the circle at the other end is empty.

The first player to get his or her three coins in a straight line wins.

1. After playing this game a few times you will note that whoever starts the game can win by following the correct strategy. What is that strategy?
2. Add the following rule: "The player who starts is not allowed to place a coin in the center circle." (Usually, the second player will fill this circle.)

This leaves the first player with six strategies:

a. Place the first two coins on opposing middle circles.
b. Place them on circles that are in the middle of segments but not opposite one another.
c. Place them on a corner and on an opposing middle circle.
d. Place them on a corner and on the middle of an adjacent segment.
e. Place them on two opposing corners.
f. Place them on two adjacent corners.

Which do you think is the best strategy?

3. Cannibal, yes or no?

THE ISLAND OF WHATIZIT IS INHABITED BY TWO TRIBES WHOSE MEMBERS LOOK ALIKE BUT DIFFER TOTALLY IN ONE RESPECT. THE MEMBERS OF ONE TRIBE, CALLED TRUTHERS, ALWAYS TELLS THE TRUTH WHILE THE MEMBERS OF THE OTHER TRIBE, CALLLED FIBBIES, ALWAYS LIE. WHEN I LAND ON THE ISLAND THREE NATIVES APPROACH ME. I HAVE NO IDEA WHICH TRIBE EACH IS FROM.

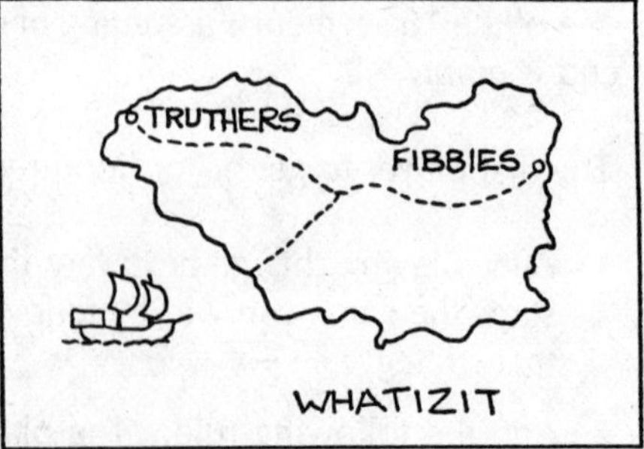

4. Dizzy diagnoses

5. Grandmother talk

LOOK OUT!

EACH GRANDMOTHER HAS TOLD THE TRUTH IN TWO OF HER THREE PRONOUNCEMENTS BUT HAS LIED IN THE THIRD. HOW MANY GRANDCHILDREN DOES EACH GRANDMOTHER ACTUALLY HAVE?

6. A worldly grandfather

7. A healthy way to use cigarettes

The ancient game of Nim is played as follows:

Open a pack of cigarettes. Divide the cigarettes into two unequal piles (for instance, 7 and 13). Choose an opponent as intelligent as yourself and take turns either removing any number of cigarettes from one pile or removing an equal number of cigarettes from both piles (for instance, if you are the first to play, you can remove 4 cigarettes from the first pile or 5 from both piles). Whoever removes the last cigarette wins. Can you suggest some guidelines for would-be victors?

Well played

1. Leapfrog

Note that at each stage of the game there cannot be more than one way to play a given color. Each game can therefore be simply described by the ordered succession of the colors used. The solution is therefore the following:

Play Black once
then White twice
then Black three times
then White three times
then Black three times
then White twice
and Black once.

There is a symmetrical solution obtained by interchanging Black and White in the list above.

General solution:

Consider a line of $(2n + 1)$ squares with n black counters on the left and n white counters on the right. Playing the game with the same rules as before, our sequence will be:

Play Black once
White twice

Black three times; and so on
Black or White $(n-1)$ times
White or Black n times
Black or White n times
White or Black n times
White or Black $(n-1)$ times; and so on
Black three times
White twice
and Black once.

2. *Game of the three roads*

Number the circles as follows:

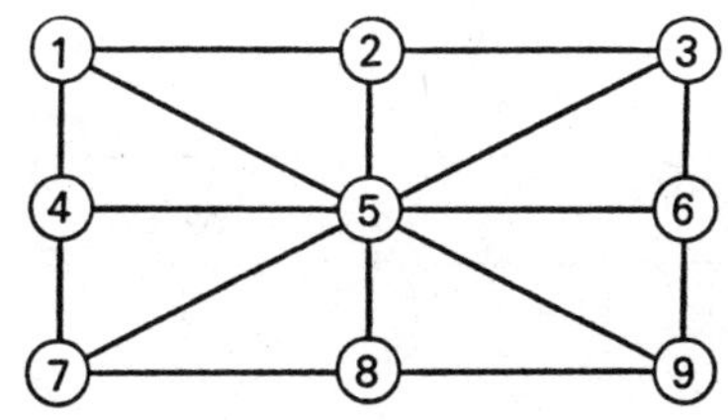

Let S be the player who starts and F be the player who finishes placing the coins.

1. The player who starts places his or her first coin at the center.

First case: F places his or her first coin on a corner. This is how the game will progress:

S	5	8	3	89	56	Wins
F	1	2	7	—	—	

Initial placement of the 3 coins — Moves from 8 to 9, 5 to 6, etc.

Second case: F places his or her first coin in the middle of a segment:

S	5	1	7	12	23	Wins
F	8	9	4	—	—	

In either case, S wins.

2. The player who starts is not allowed to place a coin at the center.

First possibility:

S	2	8	1	—	—
F	5	9	4	96	Wins

Second possibility:

S	2	6	3	—	—
F	5	7	9	58	Wins

Third possibility:

S	1	6	3	—	—
F	5	7	2	78	Wins

Fourth possibility:

S	1	2	7	—	—
F	5	3	9	56	Wins

Fifth possibility:

S	1	9	2	—	—
F	5	8	3	87	Wins

Sixth possibility:

S	1	3	8	Match
F	5	2		Draw

Strategy number six, placing the two first coins in adjacent corners, is best for player *S*: It is the only way to avoid losing (assuming that *F* plays correctly).

3. Cannibal, yes or no?

If Two Feather is telling the truth, Three Feather is lying.

If One Feather in fact said: "There is only one Truther among us," this would be wrong if he were telling the truth and correct if he were telling a lie. This is impossible, so One Feather could not have thus spoken.

Two Feather is therefore lying and Three Feather is telling the truth. I'd better spend the night in the hut of the latter.

4. Dizzy diagnoses

A careful examination of the statements made by the three consultants reveals that only one solution is possible:

I have poor eye coordination, I smoke too much, I don't eat enough chocolate, and I don't have yellow fever.

The correct statements made by the consultants are: second one for the nutritionist and first one for both the eye specialist and the family doctor.

5. Grandmother talk

There are three possible cases:

1. Emily has seven grandchildren.
Lucy is lying with her third pronouncement. She is therefore telling the truth with her other two. Contradictions are numerous in Gertrude's statements.

2. Emily has neither seven nor eight grandchildren.
Emily and Lucy are lying then when they say that she has seven or eight. Their other statements are correct but there are too many contradictions.

3. Emily has eight grandchildren.
She lies, therefore, when she says that she has seven grandchildren and she tells the truth in her other two statements. Lucy has seven grandchildren and Gertrude has ten. There are no contradictions.

6. *A worldly grandfather*

From the statement made by Cousin III, we can conclude that grandfather did not retire in Zanzibar. Cousin II is therefore telling the truth with his first statement: Grandfather was born in Zagreb and died in Zurich. The first statement of Cousin I is therefore untrue. The second is true: Grandfather did not live in Zaire.

Note once again what Cousin III said: The first statement is incorrect, therefore the second is true: Grandfather lived in Zermatt.

To summarize: Grandfather was born in Zagreb, lived in Zermatt, did not retire to Zanzibar, and died in Zurich.

7. *A healthy way to use cigarettes*

If we represent each state of the game by the number of cigarettes in the two piles, we obtain the diagram shown on the next page.

When playing the game, the state changes from move to move, either along a parallel to one of the axes or along a line at 45°. The winner is the one who reaches the origin S_0. The loser is therefore the one who ends up on one of the axes or on the first bisector. Therefore, any player who reaches either S_1 or S_1' has won because there is nowhere for the other player to go from S_1 or S_1' except to a losing state.

Working backward, any player landing on any of the three straight lines shown drawn from S_1 or S_1' will lose. Therefore, any player landing on S_2 or S_2' will win, and so on.

By similar successive steps one can identify the following states that a player must try to reach and adhere to in order to win the game:

S_1	S_2	S_3	S_4	S_5	S_6	etc.
1	3	4	6	8	9	
2	5	7	10	13	15	

and the corresponding symetrical states, as shown on the following page.

Note that the difference between the two coordinates is equal to the suffix of the corresponding state. Note also that if α is the famous golden number [root of the equation $\alpha^2 = \alpha + 1$ or $(1 + \sqrt{5})/2$ or 1.618], the coordinates of the winning states of suffix n are:

Whole part of $(n \cdot \alpha)$
Whole part of $(n \cdot \alpha) + n$

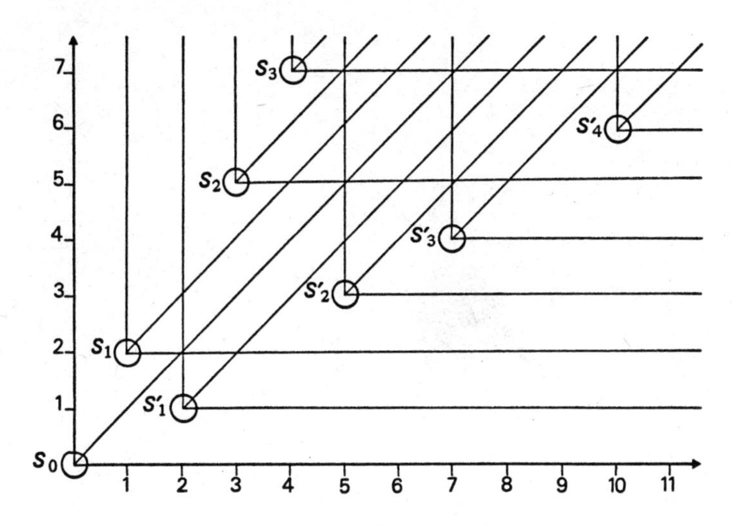

(This example is drawn from the course on game theory given by Professor Jean Bouzitat.)